Forgotten Algebra

A Self-Teaching Refresher Course

(with the optional use of the graphing calculator)

Fourth Edition

Barbara Lee Bleau, Ph.D.
Special Assistant to the Vice President
and
Professor of Mathematics
Florida Keys Community College
Key West, Florida

Meg Clemens
National Board Certified Teacher
Mathematics Department Chair and Instructor
Hugh C. Williams High School
Canton, NY

Glenn Clemens
Mathematics Instructor
Norwood-Norfolk Junior-Senior High School
Norwood, NY

BARRON'S

All inquiries should be addressed to:
Barron's Educational Series, Inc.
250 Wireless Boulevard
Hauppauge, New York 11788
www.barronseduc.com

Library of Congress Control Number: 2013935144

ISBN: 978-1-4380-0150-0

PRINTED IN THE UNITED STATES OF AMERICA
9 8 7 6 5 4 3 2 1

Contents

Preface

Algebra is not difficult! It is the primary tool used in a variety of quantitative courses in mathematics, economics, and engineering. This teach-yourself text-workbook was designed to provide students regardless of their previous background with the means to attain or improve their proficiency in this fundamental skill.

Forgotten Algebra can supplement various traditional mathematics courses, or it can be used to "brush up" before entering a course or preparing for a standardized entrance examination such as the SAT, GRE, or GMAT. It can also serve as an excellent introduction for the student who has never studied algebra. For example, each unit provides comprehensive explanations plus numerous examples, problems, and exercises, which include detailed solutions to facilitate self-study. In addition, this edition has incorporated an optional section at the end of many units entitled "Algebra and the Calculator." These sections illustrate the use of graphing calculators but are not designed to replace the manual exercises. Explanations are provided for using the Texas Instruments TI-84 model. Be circumspect in its use, however, since your time and energy are best served learning algebra rather than learning how to use a graphing calculator.

Remember, *Forgotten Algebra* is a fundamental self-help approach to providing a workbook that is easy to read yet a comprehensive guide to learning algebra.

January 2003   Barbara Lee Bleau McKinley, Ph.D.

Our goal in updating this classic book is to preserve the content and style of the original version while including more explanations to help an independent learner. This new version now has complete solutions to each problem at the end of each chapter. Complete solutions to all exercises are included at the back of the book. Graphing calculator explanations have been updated to include screenshots to match your calculator. Some additional sections and a new unit have been added to include commonly taught topics.

September 2012   Meg and Glenn Clemens

UNIT 1

Signed Numbers

In this first unit of your workbook you will learn the meanings of the terms *signed number* and *absolute value*, and the rules for performing the four basic operations of addition, subtraction, multiplication, and division using signed numbers, and how to evaluate numerical expressions using order of operations.

We will start with the concept of a signed number. A **signed number** is a number preceded by a plus or minus sign. A number preceded by a minus sign is called a **negative number**. For example, -13 and $-\dfrac{2}{5}$ are negative numbers. A plus sign is used to denote a **positive number**; $+5$ is a positive number. If no sign is written, the number is understood to be positive; 20 is a positive number. Zero is the only exception. The number 0 is neither positive nor negative.

On a number line (think of a thermometer on its side), all the numbers to the right of 0 are positive numbers and all the numbers to the left of 0 are negative numbers. The number 0, often referred to as the origin, is neither positive nor negative.

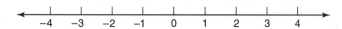

Next is the concept of absolute value. The definition that we especially like is that the **absolute value** of a signed number is the number that remains when the sign is removed. For example, the absolute value of -14 is 14; when the negative sign is removed, the number that remains is 14. Likewise, the absolute value of $+2$ is 2.

Instead of writing the words *absolute value*, we can use the symbol $| \, |$. Thus, $|-14|$ is read as "the absolute value of negative 14." And $|-14| = 14$.

Geometrically absolute value can be defined as the distance on the number line from a number to 0 without regard to direction. For example, the distance on the number line from 4 to 0 is 4 units, and the distance from -4 to 0 is 4 units. Thus, $|4| = 4$ and $|-4| = 4$.

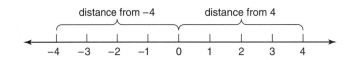

Before continuing, here are a few problems for you to try.

Problem 1   Find the absolute value of −37.

Problem 2   Find |+5|.

Problem 3   Find |−11|.

Problem 4   Find |0|.

ADDING SIGNED NUMBERS

Rule 1:   **To add numbers with like signs, add their absolute values. To this result, keep the common sign.**

EXAMPLE 1   (−5) + (−3)

Solution: The absolute value of (−5) is 5, or |−5| = 5, and the absolute value of (−3) is 3, or |−3| = 3. Add the absolute values 5 and 3. To the result, 8, keep the common sign. Thus (−5) + (−3) = −8.

EXAMPLE 2   (+5) + (+3)

Solution: Use the same reasoning as for Example 1, but the common sign now is plus. Thus, (+5) + (+3) = +8.

Rule 2:   **To add two numbers with unlike signs, subtract the smaller absolute value from the other. To this result, keep the sign of the number having the larger absolute value.**

EXAMPLE 3 $(-5) + (+3)$

Solution: $|-5| = 5$, and $|+3| = 3$. Subtract the smaller absolute value, 3, from the larger absolute value, 5. To the result, 2, keep the sign of the number having the larger absolute value, in this example, 5. Thus $(-5) + (+3) = -2$.

Now you try a few.

Problem   5 $(+5) + (+7)$

Problem   6 $(-9) + (-2)$

Problem   7 $(-1) + (-2) + (-3)$

Problem   8 $(-10) + (+2)$

Problem   9 $(7) + (-3)$

Problem 10 $(-2) + (+8)$

Problem 11 $-1 + 7$

Problem 12 $12 + (-1)$

We hope you noticed that, when adding two signed numbers, you add absolute values for like signs and subtract absolute values for unlike signs.

We fully realize you could have gotten all of Problems 5–12 correct, without ever knowing the two rules, by simply using a calculator. And if the numbers had been more complicated, say $(-1237.15) + (352.79)$, we expect that's what we would have done—used our calculator. But for easy numbers, like those above, it's faster to do them in your head than to rely on a calculator.

SUBTRACTING SIGNED NUMBERS

Rule 3:   **To subtract a signed number, add its opposite.**

EXAMPLE 4 $(5) - (-2)$

Solution: To subtract -2 from 5, the rule says to add the opposite of -2. The opposite of -2 is 2.
Thus, $(5) - (-2) = 5 + 2 = 7$.

EXAMPLE 5 $(-6) - (-2)$

Solution: The opposite of -2 is 2.
Thus, $(-6) - (-2) = -6 + 2 = -4$.

EXAMPLE 6 $(-3) - 6$

Solution: The opposite of 6 is -6.
Thus, $(-3) - 6 = (-3) + (-6) = -9$.

In each of the above examples, the subtraction was changed to addition of the opposite of the second number. Then the final answer was found by using the rules for the addition of signed numbers.

Here are some problems for you. Try them without using a calculator.

Problem 13 $23 - 18$

Problem 14 $5 - (17)$

Problem 15 $(-3) - (2)$

Problem 16 $2 - (-7)$

Problem 17 $-10 - (-5)$

Problem 18 $-11 - 2$

MULTIPLYING SIGNED NUMBERS

Rule 4:   **To multiply two signed numbers with like signs, multiply their absolute values and make the product positive.**

Rule 5:   **To multiply two signed numbers with unlike signs, multiply their absolute values and make the product negative.**

EXAMPLE   7 $(5)(2) = 10$

EXAMPLE   8 $(-3)(-5) = +15$

EXAMPLE   9 $(-1)(+7) = -7$

EXAMPLE 10 $(+3)(-2) = -6$

Think of it like this: Two like signs yield plus, and two unlike signs yield minus. Now try the following problems.

Problem 19 $(-3)(5)$

Problem 20 $(-2)(-2)$

Problem 21 $(-4)(-11)$

Problem 22 $6(-8)$

Problem 23 $(-1)(7)$

Problem 24 $(4)(-3)$

DIVIDING SIGNED NUMBERS

Rule 6:   **To divide two signed numbers with like signs, divide their absolute values and make the quotient positive.**

Rule 7:   **To divide two signed numbers with unlike signs, divide their absolute values and make the quotient negative.**

EXAMPLE 11 $-35/5 = -7$

EXAMPLE 12 $-63/-7 = 9$

EXAMPLE 13 $10/-5 = -2$

EXAMPLE 14 $-12/-2 = 6$

ORDER OF OPERATIONS

By now you should be asking yourself, what happens when the problem involves more than one operation? In such cases we use the following order of operations.

Step 1:   Perform any operations inside parentheses first. Use the order below, if necessary.

Step 2:   Working from left to right, do any multiplications or divisions in the order in which they occur.

Step 3:   Again working from left to right, do any additions or subtractions in the order in which they occur.

EXAMPLE 15

Calculate: $3 + 8 \div 4$.

Solution: $3 + 8 \div 4 = 3 + \mathbf{2}$ Divide
$\qquad\qquad\qquad = 5$ Add

EXAMPLE 16

Calculate: $16 \div 2 \cdot 4 - 6 \cdot 2$.

Solution: $16 \div 2 \cdot 4 - 6 \cdot 2 = \mathbf{8} \cdot 4 - 6 \cdot 2$ Divide
$\qquad\qquad\qquad\qquad = \mathbf{32} - \mathbf{12}$ Multiply
$\qquad\qquad\qquad\qquad = 20$ Subtract

In each of the above examples, we started on the left and did the multiplications and divisions in the order they occurred. Then, returning to the left, all the additions and subtractions were calculated in the order they occurred.

As problems increase in complexity, we will add additional steps; but for now practice using these two.

Problem 25   Calculate: $5 + 2 \cdot 4 - 3$.

Problem 26   Calculate: $27 \div 9 - 2 \cdot 7$.

EXAMPLE 17

Calculate: $2(1 - 3) - 4(5 + 2)$

Solution: $2(1 - 3) - 4(5 + 2)$ Simplify inside parentheses
$\qquad\qquad 2(-2) - 4(7)$ Multiply
$\qquad\qquad\qquad -4 - 28$ Subtract
$\qquad\qquad\qquad\qquad -32$

Here are some problems for you to try.

Problem 27   Calculate: $-3(-2 - 5) + 4(1 - 2)$

Problem 28   Calculate: $5(-6 - 1) - 4(-5 + 7)$

Algebra is an essential branch of mathematics. As such, it is important that you learn to do the math manually, that is, with paper and pencil. Notwithstanding, graphing calculators have become a way of life for some people, and we want to suggest that we take advantage of the technology available to us. All the material in this text is designed to be done manually. But at the end of some units there is an optional section entitled "Algebra and the Calculator." Explanations are provided for using a TI-84 graphing calculator. We use ours mostly when the numerical calculations are too "messy" to be done by hand and for checking our work. Optional exercises for practice and solutions are included as well. Remember, though, we want you to spend your time and energy learning algebra, which is the focus of this book, rather than learning how to use a graphing calculator.

ALGEBRA AND THE CALCULATOR (Optional)

By now maybe you've decided that a calculator is the way to go rather than learning all these rules. In each of the following examples, I have shown the keystrokes necessary to solve the problem with a calculator. Until you are more familiar with the rules of algebra it is important to enter the problem exactly as shown, especially the parentheses and signs.

EXAMPLE 18   Using a calculator, find: $-2 + (-9)$.

- -

Solution: To find the answer, press $\boxed{(-)}$ $\boxed{2}$ $\boxed{+}$ $\boxed{(}$ $\boxed{(-)}$ $\boxed{9}$ $\boxed{)}$ $\boxed{\text{ENTER}}$. Note: The $\boxed{(-)}$ key is used to denote a negative number. On your calculator, this key sequence should look like the first expression in the screen on the left below. If you pressed the subtraction key instead of the negative symbol, the screen would look like the second expression. Notice the small difference in appearance between the negative symbol and the subtraction sign. An error message, shown in the right screen below, will appear if the subtraction sign is used by mistake.

```
-2+(-9)                        ERR:SYNTAX
                          -11  1:Quit
-2+(-9)█                        2:Goto
```

Answer: -11

- -

EXAMPLE 19 Using a calculator, find: $|-6| + 3$.

Solution: Press MATH , select NUM, select 1:abs, and press ENTER . Press (−) 6 + 3 ENTER .

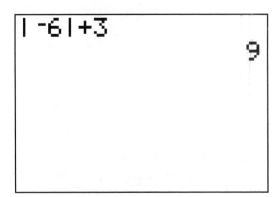

Notice the number of keystrokes needed to do this problem. You will save yourself a great deal of time if you learn to recognize problems like this as nothing more than asking for the sum: $6 + 3 = 9$.

Answer: 9

You should now have an understanding of signed numbers, absolute value, and the order of operations, and be able to add, subtract, multiply, and divide signed numbers. Briefly stated, the rules are as follows:

Addition: When the signs are the same, add the absolute values and keep the common sign. With unlike signs, subtract the absolute values and keep the sign of the number with the larger absolute value.

Subtraction: Change subtraction to addition of the opposite, and follow the rules for addition.

Multiplication and division: When the signs are the same, multiply or divide the absolute values and make the answer positive. With unlike signs, multiply or divide the absolute values and make the answer negative.

Before beginning the next unit, try the following exercises without using a calculator. When you have finished, check your answers against those at the back of the book.

EXERCISES

Perform each of the indicated operations.

1. $-2 + (-5)$
2. $1 - (-3)$

3. $8 - 17$

4. $-4 - (-3)$

5. $5(-9)$

6. $(-3)(6)$

7. $-3(-8)$

8. $32/-8$

9. $-25/-5$

10. $-3 + |-4|$

11. $|-6| - 7$

12. $90/-30$

13. $-11 + 20$

14. $-6 - (-3)$

15. $-3 + 2 + (-5)$

16. $-5 - 5$

17. $(-2)(-3)(-4)$

18. $(-4)(-8)$

19. $-3 + 1 - (-2)$

20. $(-1)(-1)(-1)(-1)$

21. $1 + 6 \cdot 3 \div 2 - 2$

22. $12 \div 3 \cdot 2 - 1$

23. $0 \cdot 7 + 15 \div 5 - 2 \cdot 1$

24. $(-10) \cdot 12 - 6 \cdot 3$

25. $|-5| + |3| - |-2|$

26. $-|-3| + |7|$

27. $-6(7 - 4) + 2(3 - 5)$

28. $7(3 - 6) - 8(4 - 1)$

Solutions to Problems

1. $|-37| = 37$

2. $|+5| = 5$

3. $|-11| = 11$

4. $|0| = 0$

5. $(+5) + (+7) = 12$. The signs are the same. Add the values and keep the sign.

6. $(-9) + (-2) = -11$. The signs are the same. Add the values and keep the sign.

7. $(-1) + (-2) + (-3) = -6$. The signs are the same. Add the values and keep the sign.

8. $(-10) + (+2) = -8$. The signs are different so subtract the values. $10 - 2 = 8$. Keep the negative sign from the larger value, -10.

9. $(7) + (-3) = 4$. The signs are different so subtract the values. $7 - 3 = 4$. Keep the positive sign from the larger value, $+7$.

10. $(-2) + (+8) = 6$. The signs are different so subtract the values. $8 - 2 = 6$. Keep the positive sign from the larger value, $+8$.

11. $-1 + 7 = 6$. The signs are different so subtract the values. $7 - 1 = 6$. Keep the positive sign from the larger value, $+7$.

12. $12 + (-1) = 11$. The signs are different so subtract the values. $12 - 1 = 11$. Keep the positive sign from the larger value, $+12$.

13. $23 - 18 = 5$. No need for new rules, just subtract.

14. $5 - (17) = -12$. Change to $5 + -17$. Subtract the values and keep the negative sign from the larger value, -17.

15. $(-3) - (2) = -5$. Change to $-3 + -2$. Add the values and keep the common sign.

16. $2 - (-7) = 9$. Change to $2 + (+7) = 2 + 7 = 9$.

17. $-10 - (-5) = -5$. Change to $-10 + (+5) = -10 + 5$. Subtract the values and keep the negative sign from the larger value, -10.

18. $-11 - 2 = -13$. Change to $-11 + -2$. Add the values and keep the common sign.

19. $(-3)(5) = -15$

20. $(-2)(-2) = 4$

21. $(-4)(-11) = 44$

22. $6(-8) = -48$

23. $(-1)(7) = -7$

24. $(4)(-3) = -12$

25. $\begin{aligned} 5 + 2 \cdot 4 - 3 &= 5 + 8 - 3 \\ &= 13 - 3 \\ &= 10 \end{aligned}$

26. $\begin{aligned} 27 \div 9 - 2 \cdot 7 &= 3 - 2 \cdot 7 \\ &= 3 - 14 \\ &= -11 \end{aligned}$

27. $\begin{aligned} -3(-2 - 5) + 4(1 - 2) &= -3(-7) + 4(-1) \\ &= 21 + -4 \\ &= 17 \end{aligned}$

28. $\begin{aligned} 5(-6 - 1) - 4(-5 + 7) &= 5(-7) - 4(2) \\ &= -35 - 8 \\ &= -43 \end{aligned}$

UNIT 2

Grouping Symbols and Their Removal

In this unit you will learn the meanings of some of the words we will be using throughout the course. Then you will learn about various grouping symbols and the way they are removed to simplify expressions.

SOME IMPORTANT DEFINITIONS

Five of the most important concepts you will need in the course are: *variable, term, coefficient, expression*, and *like terms*.

Here are some "commonsense" or "working" definitions of these concepts. It is *not* important that you memorize these definitions, but you should understand each one and be able to define the concept in your own words.

A **variable** is a letter or symbol used to represent some unknown quantity. For example, x is often used as a variable.

A **term** is a symbol or group of symbols separated from other symbols by a plus or minus sign. For example, in $3x - 5y + 7xyz - 2$, there are four terms. Note that terms can contain a number and/or several letters. Therefore $7xyz$ is a single term; the number is written first, and the letters are usually put in alphabetical order.

The numerical **coefficient** of a variable is the number that multiplies the variable. In the term $2y$ the coefficient of y is 2. Here are some examples:

a. $8w$ The numerical coefficient of w is 8.

b. x When no number is written, the numerical coefficient is understood to be 1.

c. $-5z$ The numerical coefficient is -5.

The sum or difference of one or more terms can be referred to as an **expression**. Expressions are given different names depending on the number of terms involved. Expressions with one term are called monomials, whereas expressions with two terms are called binomials. The expression $3x - 5y + 4w$ has three terms and is called a trinomial.

Like terms contain the same variable or variables and differ only in their numerical coefficients. In the expression $3x + 4x$ the terms are like terms that differ only with respect to their coefficients, 3 and 4. Like terms can be combined. The terms of the expression $3x + 4x$ can be combined to form a single term, $7x$. Similarly, like terms in the trinomial expression $3x - 5y + 4x$ can be combined, forming a binomial expression, $7x - 5y$.

Try a few problems to test your understanding of these important concepts.

Problem 1   How many terms are there? $10zy + 3x - 5an + 12$

Problem 2   Which is the *variable*? $7a$

Problem 3   What is the *coefficient* of y? $5y$

Problem 4   A binomial expression has _____ terms.

Problem 5   When the like terms in the expression $10y + 8z + 3y$ are combined, the result will be _____.

GROUPING SYMBOLS

Now let's turn our attention to grouping symbols. There are two common types of grouping symbols:

parentheses   ( )
brackets   [ ]

Whether you use parentheses or brackets, these symbols indicate that the quantities enclosed within them are considered to be a single unit with respect to anything outside the grouping symbol. In the expression $8y + (3a - b) + 2x$, the terms $3a - b$ are considered as a single unit, separate from the terms outside the parentheses.

REMOVING GROUPING SYMBOLS AND SIMPLIFYING

To remove grouping symbols we use the **distributive property:** $a(b + c) = ab + ac$. The distributive property tells us to remove parentheses by multiplying each term inside the parentheses by the term in front of the parentheses.

REMOVING GROUPING SYMBOLS AND SIMPLIFYING

EXAMPLE 1 $2(x + y) = 2x + 2y$

In an expression like $2(x + y)$, we multiply each term inside the parentheses by 2.

EXAMPLE 2 $-3(x - 5) = -3x + 15$

Note the plus sign between the terms since $(-3)(-5) = +15$.

Next we consider some more complicated expressions involving like terms. To **simplify** these expressions, we follow the rules of order of operations from Unit 1. Working from left to right, first remove the grouping symbols (step 1, do any multiplications or divisions) followed by combining any like terms (step 2, do any additions and subtractions).

EXAMPLE 3

Simplify: $2(4xy + 3z) + 3(x - 2xy) - 4(z - 2xy)$.

Solution: $2(4xy + 3z) + 3(x - 2xy) - 4(z - 2xy)$

$= 8xy + 6z + 3x - 6xy - 4z + 8xy$

Note plus sign.

$= 3x + 2z + 10xy$

EXAMPLE 4

Simplify: $3aw + (aw + 4z) - (2x - 3y + 4z)$.

Solution: $3aw + (aw + 4z) - (2x - 3y + 4z)$

Note: To remove the first set of parentheses we are actually multiplying by 1, which is understood to be there but is not written. In an expression like $-(2x - 3y + 4z)$, remove the parentheses by multiplying each term inside the parentheses by -1. Again the 1 is not written, but it is understood to be there. In this example, we will rewrite the expression with 1s to be clear about the distributive property.

$= 3aw + 1(aw + 4z) - 1(2x - 3y + 4z)$

$= 3aw + aw + 4z - 2x + 3y - 4z$

$= 4aw - 2x + 3y$

Try the next two on your own.

Problem 6

Remove the grouping symbols and simplify: $3(x - 10 + 2y) - 2$.

Problem 7

Simplify: $2(x - 1) - 3(2 - 3x) - (x + 1)$.

Solution:

Occasionally more complicated expressions occur with parentheses within parentheses. When one set of symbols is within another, the grouping symbols must be removed from the inside out. One example follows.

EXAMPLE 5

Simplify: $x - [5 - 2(x - 1)]$.

Solution: $x - [5 - 2(x - 1)]$

$$= x - [5 - 2x + 2]$$
$$= x - 1[5 - 2x + 2]$$

$$= x - 5 + 2x - 2$$
$$= 3x - 7$$

ALGEBRA AND THE CALCULATOR (Optional)

Be careful when evaluating expressions with multiple grouping symbols on the graphing calculator. On a graphing calculator, only use parentheses ( ) for grouping; do not use brackets [ ] or { }. Brackets have a different mathematical meaning than parentheses.

EXAMPLE 6

Evaluate: $x - [5 - 2(x - 1)]$ for $x = 10$

Solution: Substitute $x = 10$ and use a calculator to evaluate: $10 - [5 - 2(10 - 1)]$. Use only ( ) symbols.

Correct

```
10-(5-2(10-1))
              23
■
```

Wrong

```
10-[5-2(10-1)]
           Error
```

The Table feature of a graphing calculator can be used to compare different forms of an expression. This is a particularly useful skill on multiple choice exams.

EXAMPLE 7

Which choice is equivalent to $5 - 2(x - 1)$?

   (a) $3x - 3$ (b) $-2x + 3$ (c) $-2x + 7$

Solution: $5 - 2(x - 1)$

            $5 - 2x + 2$ Distribute

            $-2x + 7$ Combine like terms

Use a graphing calculator to check each choice. Press the Y=key, $\boxed{Y=}$, and type the original expression in the first line. To type the variable x, use the $\boxed{X,T,\theta,n}$ key.

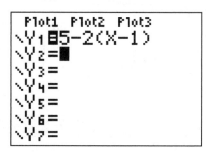

```
Plot1 Plot2 Plot3
\Y1=5-2(X-1)
\Y2=■
\Y3=
\Y4=
\Y5=
\Y6=
\Y7=
```

Type the expression for each choice, one at a time, in the second line. Here is choice (a).

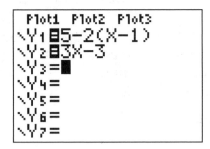

Now look at the table of values for these expressions. If the expressions are equivalent, they will have the same *y*-values in the table for each *x*-value. To display the table, press 2nd Graph, [2nd] [GRAPH].

X	Y₁	Y₂
0	7	-3
1	5	0
2	3	3
3	1	6
4	-1	9
5	-3	12
6	-5	15

Press + for △Tbl

Compare the values in the column for Y1 and Y2. If they are not equal, then the expressions are not equivalent.

Press Y= and clear the second line by pressing the down arrow and then the CLEAR key. Type the expression for choice (b) in Y2 and check the table.

Plot1 Plot2 Plot3
\Y₁■5−2(X−1)
\Y₂■ -2X+3
\Y₃=█
\Y₄=
\Y₅=
\Y₆=
\Y₇=

X	Y₁	Y₂
0	7	3
1	5	1
2	3	-1
3	1	-3
4	-1	-5
5	-3	-7
6	-5	-9

Press + for △Tbl

The values in the columns for Y1 and Y2 are not equal. Clear Y2 and type in the expression for choice (c).

Plot1 Plot2 Plot3
\Y₁■5−2(X−1)
\Y₂■ -2X+7
\Y₃=█
\Y₄=
\Y₅=
\Y₆=
\Y₇=

X	Y₁	Y₂
0	7	7
1	5	5
2	3	3
3	1	1
4	-1	-1
5	-3	-3
6	-5	-5

Press + for △Tbl

Now the values for Y1 and Y2 are equal. Choice (c) is equivalent to the original expression. Even if a question is not multiple choice, this method can be used to check if your simplified answer is equivalent to the original expression.

You should now be able to define, in your own words, the basic concepts *term*, *variable*, *coefficient*, *expression*, and *like terms*. These basic concepts will be used continually throughout the following units.

You should also be familiar with three common types of grouping symbols. Remember that the quantities enclosed within these symbols are considered to be a single unit, separate from anything outside the parentheses or brackets.

Remember also that, when one set of symbols is within the other, grouping symbols are removed by working from the inside out. Simplifying then involves combining the like terms.

Before beginning the next unit you should simplify the following expressions by removing the grouping symbols and collecting like terms. When you have completed them, check your answers against those at the back of the book.

EXERCISES

Simplify:

1. $-7(x + 2y - 3)$

2. $3x + 4x + (x + 2)$

3. $2(x + y) + 7x + 3$

4. $-(-2x + 1) + 1$

5. $7x - 2y + 5 + (2x + 5y - 4)$

6. $(3x + 5xy + 2y) + (4 - 3xy - 2x)$

7. $(5y - 2a + 1) + 2(y - 3a - 7)$

8. $3(2a - b) - 4(b - 2a)$

9. $2(7x - 5 + y) - (y + 7)$

10. $5 - 2[x + 2(3 + x)]$

11. $x - [2x - 2(1 - x)]$

12. $4 - 9(2x - 3) + 7(x - 1)$

13. $3(x + 4) + 5x - 8$

14. $4x - [5 - 3(2x - 6)]$

Solutions to Problems

1. four

2. a

3. 5

4. two

5. $13y + 8z$

6. $3(x - 10 + 2y) - 2 = 3x - 30 + 6y - 2$

$$= 3x + 6y - 30 - 2$$
$$= 3x + 6y - 32$$

7. $2(x - 1) - 3(2 - 3x) - (x + 1) = 2(x - 1) - 3(2 - 3x) - 1(x + 1)$

$$= 2x - 2 - 6 + 9x - x - 1$$
$$= 2x + 9x - x - 2 - 6 - 1$$
$$= 10x - 9$$

UNIT 3

Solving First-Degree Equations

The purpose of this unit is to provide you with an understanding of first-degree equations. When you have finished the unit, you will be able to identify first-degree equations, distinguish them from other types of equations, and solve them.

What is a first-degree equation? A **first-degree equation** has these characteristics:

1. There is **only one variable**.

2. The variable is involved in **one or more of only the four fundamental operations** of addition, subtraction, multiplication, and division.

3. The variable is **never multiplied by itself**.

4. The variable does **not** appear **in any denominator**.

Here are some examples of first-degree equations:

$$2x + 3 = 15$$
$$\sqrt{5}x - \frac{x+2}{2} = \pi$$
$$3(x-1) + 2 = 0$$

Here are some examples that are *not* first-degree equations:

$$x + 3y = 5$$
$$x^2 = 9$$
$$3 + \sqrt{x} = 2(x-7)$$
$$\frac{2}{x} = 7(x-1)$$
$$(x-3)(x+3) = 12$$

Before proceeding, determine why each of the above is *not* a first-degree equation.

Now we are ready to begin solving first-degree equations. You probably remember the basic strategy—isolate the variable on one side of the equal sign and get all other terms on the other side. To accomplish this, we use two rules. The first is:

Rule 1: **A term can be moved from one side of an equation to the other by adding its opposite to both sides of the equation.**

When we move a negative term, we sometimes say that we "add the same number on both sides of the equation." Likewise, when we move a positive term, we can say that we "subtract the same number on both sides of the equation."

EXAMPLE 1

$$2x - 3 = 5 + x$$

$\underline{-x \qquad\quad -x}$ Add the opposite of $+x$ to both sides.

$x - 3 = 5$ Combine terms.

$\underline{\quad +3\ +3\quad}$ Add the opposite of -3 to both sides.

$x = 8$ Combine terms.

Rule 1 Shortcut: **A term can be transposed or moved from one side of an equation to the other if and only if its sign is changed to its opposite as it crosses the equal sign.**

EXAMPLE 2

If $2 - x = 7,$

then $2(-x) = (7)$

$2 - 7 = x,$

and $-5 = x.$ Note that the signs have changed.

The second rule is:

Rule 2: **Both sides of an equation can be multiplied or divided by the same nonzero number.**

EXAMPLE 3

If $3x = 2,$

then $\dfrac{3x}{3} = \dfrac{2}{3},$

$\dfrac{3x}{3} = \dfrac{2}{3},$

and $x = \dfrac{2}{3}.$

EXAMPLE 4   If $\dfrac{x}{4} = 12,$

then $4\left(\dfrac{x}{4}\right) = 4(12),$

$\dfrac{4x}{4} = 48,$

and $x = 48.$

Now let's get on with the business of solving equations.

> Definition: A **solution** of an equation is any number that makes the equation true when that number is substituted for the variable. Sometimes it is called the **root** of the equation.

To solve first-degree equations, I use a four-step strategy that can be used to solve *all* first-degree equations. The four steps are as follows:

> **Strategy for Solving First-Degree Equations**
>
> 1. **Simplify:**
>   a. Clear any fractions.
>   b. Remove parentheses.
>   c. Collect like terms.
>
> 2. **Transpose:**
>   Isolate all terms with the variable to one side and transpose everything else to the other side. Remember to change the term's sign when crossing the equal sign.
>
> 3. **Simplify.**
>
> 4. **Divide** each term of the entire equation **by the coefficient** of the variable.

Here are some examples that illustrate the use of these steps in solving first-degree equations. Read through each step, and be sure you understand what has happened to the terms.

EXAMPLE 5

Solve the equation for x: $5x + 10 - 3x = 6 - 4x + 16.$

Solution: $5x + 10 - 3x = 6 - 4x + 16$

1. **Simplify** by collecting like terms. $2x + 10 = 22 - 4x$

2. **Transpose.** $2 + 10 = 22 - 4x$

$2x + 4x = 22 - 10$

3. **Simplify** by collecting like terms. $6x = 12$

4. **Divide** by the coefficient of x. $\dfrac{6x}{6} = \dfrac{12}{6}$

$$\dfrac{\cancel{6}x}{\cancel{6}} = 2$$

Answer: $x = 2$

Check: Always check your answer in the original equation using order of operations.

$$5x + 10 - 3x = 6 - 4x + 16$$
$$5(2) + 10 - 3(2) = 6 - 4(2) + 16$$
$$10 + 10 - 6 = 6 - 8 + 16$$
$$14 = 14$$

EXAMPLE 6

Solve the equation for x: $3(2x + 5) = 4x + 23$.

Solution: $3(2x + 5) = 4x + 23$

1. **Simplify** by removing parentheses. $6x + 15 = 4x + 23$

2. **Transpose.** $6x + 15 = 4x + 23$

$$6x - 4x = 23 - 15$$

3. **Simplify** by collecting like terms. $2x = 8$

4. **Divide** by the coefficient of x. $\dfrac{\cancel{2}x}{\cancel{2}} = \dfrac{8}{2}$

Answer: $x = 4$

Check:
$$3(2x + 5) = 4x + 23$$
$$3(2(4) + 5) = 4(4) + 23$$
$$3(8 + 5) = 16 + 23$$
$$3(13) = 39$$
$$39 = 39$$

EXAMPLE 7

Solve the equation for x: $4 - (2x - 8) = x$.

Solution:

$$4 - (2x - 8) = x$$

1. **Simplify** by removing parentheses,
    and collecting like terms.

    $$4 - 2x + 8 = x$$
    $$12 - 2x = x$$

2. **Transpose.**

    $$12 \underset{\frown}{- 2x} = x$$
    $$12 = x + 2x$$

3. **Simplify** by collecting like terms.

    $$12 = 3x$$

4. **Divide by the coefficient.**

    $$\frac{12}{3} = \frac{3x}{3}$$

Answer:

$$4 = x$$

Try to solve this first-degree equation yourself. Cover the solution below, and refer to it only after you have arrived at *your* solution.

Problem 1

Solve the equation for x: $x - 3 - 2(6 - 2x) = 2(2x - 5)$.

1. **Simplify:** remove parentheses,   _____ = _____

    collect like terms.   _____ = _____

2. **Transpose.**   _____ = _____

3. **Simplify:** collect like terms.   _____ = _____

Solution:

Now try solving this equation without any clues.

Problem 2

Solve the equation for x: $3x - 2(x + 1) = 5x - 6$.

Solution:

- -

Many people seem to panic when fractions occur in a problem. For that reason, when attempting to solve an equation containing fractions, I suggest immediately multiplying the entire equation (Rule 2) by the common denominator to clear it of fractions. The next two examples will illustrate this concept.

==

EXAMPLE 8

Solve the equation for x: $\dfrac{3x}{2} - 5x = 6$.

Solution: $\dfrac{3x}{2} - 5x = 6$

   1. **Simplify** by clearing fractions.

     (multiply equation by the denominator) $2\left(\dfrac{3x}{2} - 5x = 6\right)$

     (this cancels the denominator) $\dfrac{\cancel{2} \cdot 3x}{\cancel{2}} - 2 \cdot 5x = 2 \cdot 6$

     (equation cleared of fraction) $3x - 10x = 12$

   3. **Simplify** by collecting like terms. $-7x = 12$

     $\dfrac{\cancel{-7}x}{\cancel{-7}} = \dfrac{12}{-7}$

   4. **Divide by the coefficient of x.**

     $x = \dfrac{12}{-7} = \dfrac{-12}{7} = -\dfrac{12}{7}$

Answer:

Note: The answer can be written in three different ways. We happen to prefer the second fraction with the negative in the numerator, but either the second or third fraction is customary.

EXAMPLE 9

Solve: $\dfrac{2}{5}x - 4 = \dfrac{1}{3}x + 1.$

Solution: $\dfrac{2}{5}x - 4 = \dfrac{1}{3}x + 1$

1.   **Simplify** by clearing fractions.

(multiply by the common denominator) $15\left(\dfrac{2}{5}x - 4 = \dfrac{1}{3}x + 1\right)$

(this cancels the denominators) $15\cdot\dfrac{2}{5}x - 15\cdot 4 = 15\cdot\dfrac{1}{3}x + 15\cdot 1$

(equation cleared of fractions) $3\cdot 2x - 60 = 5\cdot 1x + 15$

$6x - 60 = 5x + 15$

2.   **Transpose.** $6x - 60 = 5x + 15$

$6x - 5x = 15 + 60$

3.   **Simplify** by collecting like terms. $x = 75$

Answer: $x = 75$

TYPES OF EQUATIONS

Thus far, all the equations we have considered have been what are called conditional equations. A **conditional equation** is one that is true for only certain values of a variable. For example, $x + 5 = 8$ is true only for $x = 3$. Therefore $x + 5 = 8$ is a conditional equation.

Now in contrast, consider the following example.

EXAMPLE 10

Solve: $3x + 1 + x = 2(x + 1) + 2x - 1.$

Solution: $3x + 1 + x = 2(x + 1) + 2x - 1$

1.   **Simplify** by removing parentheses, $3x + 1 + x = 2x + 2 + 2x - 1$

and collecting like terms. $4x + 1 = 4x + 1$

2.   **Transpose.** $4x + 1 = 4x + 1$

$4x - 4x = 1 - 1$

3.   **Simplify** by collecting like terms. $0 = 0$

Answer: The solution is the entire set of real numbers.

Before continuing, let me make sure you understand why the solution to Example 10 is the set of real numbers. The final equation, $0 = 0$, is *always* true, regardless of the value of x, because 0 always equals 0. This type of equation is called an **identity**. Its solution is the entire set of real numbers. In other words, x can equal any number.

For example, if $x = 5$, then $3(5) + 1 + 5 = 2(5 + 1) + 2(5) - 1$,

$$15 + 1 + 5 = 12 + 10 - 1,$$

$$\text{and } 21 = 21.$$

Or,   if $x = -3$, then $3(-3) + 1 + (-3) = 2(-3 + 1) + 2(-3) - 1$,

$$-9 + 1 + (-3) = -2(2) - 6 - 1,$$

$$-11 = -4 - 6 - 1,$$

$$\text{and } -11 = -11.$$

Try *any* value you like for x. Prove to yourself that it does indeed satisfy the equation.

Before leaving this example, notice that midway through the solution the left side of the equation was simplified to $4x + 1$, as was the right side of the equation. We could have stopped there because $4x + 1 = 4x + 1$ is *always* true, regardless of the value of x.

Solve the next problem on your own.

Problem 3

Solve: $6(x + 1) - x = 2x + 3(x + 2)$

Solution:

Will there always be a solution to an equation? What do you think? The answer is no, although the majority of the problems in this book have solutions because you need practice in solving equations. The last example illustrates how you can recognize when there is no solution to an equation.

EXAMPLE 11

Solve: $3(x + 1) + x = 4(x + 1)$.

Solution:

1. **Simplify** by removing parentheses, then collecting like terms.

2. **Transpose.**

3. **Simplify.**

$$3(x + 1) + x = 4(x + 1)$$

$$3x + 3 + x = 4x + 4$$
$$4x + 3 = 4x + 4$$

$$4x + 3 = 4x + 4$$

$$4x - 4x = 4 - 3$$
$$0 = 1$$

Answer: No solution

The final equation, $0 = 1$, is *always* false, regardless of the value of x, because 0 never equals 1. No matter what value is substituted in for the variable, the final statement will be false. Thus, whenever the concluding equation is false, there is no solution to the problem.

These same four steps apply when we solve first-degree equations of greater difficulty. For example, consider the equation: $3(1 + 4x - (x + 1)) = 0$.

Note that one pair of symbols is "nested" within the outer pair. Recall from the previous unit that we start by removing the innermost pair and then work outward. We will conclude this unit by your solving this last problem.

Problem 4

Solve: $6(4x - 1) - 3(4x - 2) = 7x + 5(x + 1)$.

Solution:

ALGEBRA AND THE CALCULATOR (Optional)

You can use the store feature of a graphing calculator to quickly and accurately check your solution.

EXAMPLE 12

Solve and check: $3(2x + 5) = 4x + 23$

Solution: In Example 6, we found the solution to be $x = 4$.

Check this solution on your calculator using the store feature. First, store your solution in x by pressing 4, then the store key, $\boxed{\text{STO>}}$ (found above the ON key), which displays as an arrow, then press the key for x, $\boxed{\text{X,T,}\theta\text{,n}}$.

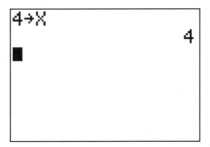

The calculator now substitutes 4 as the value for x in the next expressions you enter. Enter the left side of the original, unsimplified equation. Press enter, then enter the right side of the original equation.

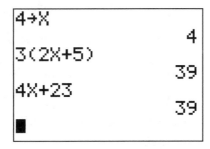

Your solution is correct if both expressions evaluate to the same number. In this example, when $x = 4$, both sides evaluate to 39. When we did the check in Example 6, we also found that both sides of the equation evaluated to 39. If you made an error in your solution, your x-value will not check.

This technique is particularly convenient for checking equations with fractions. Fractions can be entered using the fraction menu found in F1. To access the fraction menu, press ALPHA Y= .

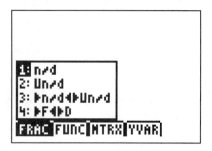

Press Enter or 1 for a stacked (up and down) fraction. Type the values in the fraction template, using the arrow keys to move to the denominator and then outside the fraction template. Here is $\frac{2}{3}$.

EXAMPLE 13

Solve and check: $\dfrac{3x}{2} - 5x = 6$

Solution: In Example 8, we found the solution to be $x = \dfrac{-12}{7}$.

To check this solution without a calculator, you would need to simplify $\dfrac{3\left(\dfrac{-12}{7}\right)}{2} - 5\left(\dfrac{-12}{7}\right)$,

which looks nasty. Here is the calculator check. First, store your solution in x. Then evaluate each side of the equation. In this example, the right side is 6, so just evaluate the left side to check if it is equal to 6.

```
-12
─── →X
 7
                    -12
                    ───
                     7
3X
── -5X
2
                      6
```

A calculator can easily do this check.

You should now be able to solve any first-degree equation. Remember that the basic strategy for solving a first-degree equation involves four steps:

1. **Simplify** by removing parentheses, clearing fractions, and then collecting like terms.

2. **Transpose.**

3. **Simplify** by collecting like terms.

4. **Divide by the coefficient of the variable.**

Also remember that the sign of a term changes when the term is moved across the equal sign, and that both sides of an equation can be multiplied or divided by the same nonzero number.

I believe this is one of the most important units of the book because first-degree equations reoccur throughout algebra. Therefore I want to encourage you to make sure you are able to successfully solve the following equations before beginning the next unit.

EXERCISES

Solve:

1. $2x - 7 = 9 - 6x$

2. $2(x + 1) - 3(4x - 2) = 6x$

3. $\dfrac{x - 3}{4} = 5$

4. $20 - \dfrac{3x}{5} = x - 12$

5. $2x - 9 = 5x - 15$

6. $2(x + 2) = 5 + \dfrac{x + 1}{3}$

7. $15 - 3(9 - x) = x$

8. $3 - \dfrac{5(x - 1)}{2} = x$

9. $x - \dfrac{x - 1}{4} = 0$

10. $1 - \dfrac{x}{2} = 5$

11. $-5w - 1 = -9w - 1$

12. $3(-2x + 1) = -6x - 7$

13. $2(4a + 1) = 4(2a - 1) + 6$

14. $3(z + 5) - 2z = \dfrac{z - 1}{2} + 17$

15. $13 - (2y + 2) = 2(y + 2) + 3y$

16. $3x + 4(x - 2) = x - 5 + 3(2x - 1)$

17. $2c + 3(c + 2) = 5c + 11$

18. $2[2 - x - (2x - 5)] = 11 - x$

Solutions to Problems

1. $x - 3 - 2(6 - 2x) = 2(2x - 5)$

$$x - 3 - 12 + 4x = 4x - 10$$

$$5x - 15 = 4x - 10$$

$$5x - 4x = -10 + 15$$

$$x = 5$$

2. $3x - 2(x + 1) = 5x - 6$

$$3x - 2x - 2 = 5x - 6$$

$$x - 2 = 5x - 6$$

$$x - 5x = -6 + 2$$

$$-4x = -4$$

$$\frac{-4x}{-4} = \frac{-4^1}{-4}$$

$$x = 1$$

3. $6(x + 1) - x = 2x + 3(x + 2)$

$$6x + 6 - x = 2x + 3x + 6$$

$$5x + 6 = 5x + 6$$

The solution is the entire set of real numbers.

4. $6(4x - 1) - 3(4x - 2) = 7x + 5(x + 1)$

$$6(4x - 1) - 3(4x - 2) = 7x + 5(x + 1)$$

$$24x - 6 - 12x + 6 = 7x + 5x + 5$$

$$12x = 12x + 5$$

$$12x - 12x = 5$$

$$0 = 5$$

The equation has no solution.

UNIT 4

A Special Type of Equation: The Fractional Equation

In this unit you will learn about a special type of equation, the fractional equation, and the strategy used to solve it. When you have completed the unit, you will be able to identify and solve fractional equations.

> Definition: A **fractional equation** is an equation in which the variable appears in a denominator.

For example, $\dfrac{2+x}{x} = 3$ is a fractional equation.

The same four steps we have been using to solve first-degree equations will continue to be used. However, a fifth step must be added when solving fractional equations. We must check to see whether our solution satisfies the **original** equation.

Recall that a solution is any number that makes the equation true when that number is substituted for the variable. Unfortunately there are several operations that may produce an equation not equivalent to the original equation. One of these operations is multiplying or dividing both sides of an equation by an expression containing the variable, which often occurs when solving a fractional equation. Thus, the final test of whether a number is part of the solution for a fractional equation is to substitute it into the original equation and see whether it yields a true statement.

> **Strategy for Solving Fractional Equations**
>
> 1. **Simplify:** clear any fractions,
>   remove parentheses,
>   collect like terms.
> 2. **Transpose.**
> 3. **Simplify.**
> 4. **Divide by the coefficient of the variable.**
> 5. **Check** by substituting the tentative answer into the original equation.

Keep in mind that as the equations become more complex, you might want to interchange the order in which you simplify. For example, it might be more logical in a given problem to remove parentheses before clearing the fractions. As long as you have carried out the algebraic procedures correctly, the resulting equations will be equivalent.

At this time it is necessary to remind you that **division by zero is undefined**. In other words, zero can never be the denominator of a fraction; $\frac{a}{0}$, for example, is meaningless.

We will now use the four basic steps and the check by substitution to solve our example of a fractional equation.

EXAMPLE 1

Solve for x: $\dfrac{2+x}{x} = 3.$

Solution: 1. **Simplify** by clearing fractions.

$$\left(\frac{2+x}{x} = 3\right)x$$

$$\frac{(2+x)}{x} \cdot x = 3 \cdot x$$

$$\frac{(2+x)}{\cancel{x}} \cdot \cancel{x} = 3 \cdot x$$

$$2 + x = 3x$$

2. **Transpose.** $2 = 3x - x$

3. **Simplify.** $2 = 2x$

4. **Divide** by the coefficient of x. $\dfrac{\cancel{2}}{\cancel{2}} = \dfrac{\cancel{2}x}{\cancel{2}}$

$$1 = x$$

5. **Check:** substitute the tentative
answer in the **original** equation.

$$\frac{2+x}{x} = 3$$

$$\frac{2+1}{1} \stackrel{?}{=} 3$$

$$3 = 3$$

Answer: This is true. So $x = 1$ is the solution.

_____

EXAMPLE 2

Solve for x: $\dfrac{7}{x} + 3 = \dfrac{-1}{2}$.

_____

Solution: What shall we use as the common denominator to clear the equation of fractions: x, $2x$, $3x$, 2, etc.?

Using our method, we can make a wrong choice and still solve the equation simply by adding an extra step or two.

Suppose we use x:

$$\frac{7}{x} + 3 = \frac{-1}{2}$$

$$\left(\frac{7}{x} + 3 = \frac{-1}{2}\right)x$$

$$\frac{7}{x} \cdot x + 3 \cdot x = \frac{-1}{2} \cdot x$$

$$7 + 3x = \frac{-x}{2}$$

Obviously x was not a wise choice as it did not clear the equation completely of fractions. However, it is easy to continue the solution.

Because there is still a denominator of 2, we can simply repeat the process using 2:

$$2\left(7 + 3x = \frac{-x}{2}\right)$$

$$2 \cdot 7 + 2 \cdot 3x = 2\left(\frac{-x}{2}\right)$$

$$14 + 6x = -x$$

$$6x + x = -14$$

$$7x = -14$$

$$\frac{7x}{7} = \frac{-14}{7}$$

$$x = -2$$

Had we multiplied by $2x$ in the first step, the problem would have been much shorter.

Now we must check the tentative solution, using the original equation.

$$\frac{7}{x} + 3 = \frac{-1}{2}$$

$$\frac{7}{-2} + 3 \overset{?}{=} \frac{-1}{2}$$

$$\frac{-7}{2} + 3 \overset{?}{=} -\frac{1}{2}$$

$$-\frac{1}{2} = -\frac{1}{2}$$

Answer: This is true. So $x = -2$ is the solution.

_____

Now you try one.

Problem 1

Solve for x: $\quad \dfrac{2}{x+1} + 3 = 1$.

Hint: $(x + 1)$ is the **entire** denominator.

_____

Solution:

_____

We assume that by now you have learned the basic strategy for solving first-degree equations, and so we have omitted writing out the directions for the first four steps in the following examples. However, we will continue to remind you of step 5, that is, to check your tentative solution. Here are two more examples.

EXAMPLE 3

Solve for x: $\quad \dfrac{2(x+1)}{x} = \dfrac{2}{x}$.

_____

Solution:

$$\frac{2(x+1)}{x} = \frac{2}{x}$$

$$x\left(\frac{2(x+1)}{x} = \frac{2}{x}\right)$$

$$\frac{2x(x+1)}{\cancel{x}} = \frac{2\cancel{x}}{\cancel{x}}$$

$$2(x+1) = 2$$

$$2x + 2 = 2$$

$$2x = 2 - 2$$

$$\frac{\cancel{2}x}{\cancel{2}} = \frac{0}{2}$$

$$x = 0$$

Check: $$\frac{2(x+1)}{x} = \frac{2}{x}$$

$$\frac{2(0+1)}{0} \overset{?}{=} \frac{2}{0}$$

However, recall that we said that **division by zero is undefined**! Therefore $x = 0$ cannot be a solution to this equation.

Answer: The equation has no solution.

- -

What Example 3 should impress on you is the necessity for checking the tentative solution.

EXAMPLE 4

Solve for x: $\dfrac{2x-4}{x-3} = 3 + \dfrac{2}{x-3}$.

- -

Solution:

$$\frac{2x-4}{x-3} = 3 + \frac{2}{x-3}$$

$$\frac{2x-4}{\cancel{x-3}}(\cancel{x-3}) = 3(x-3) + \frac{2}{\cancel{x-3}}(\cancel{x-3})$$

$$2x - 4 = 3(x-3) + 2$$

$$2x - 4 = 3x - 9 + 2$$

$$2x - 4 = 3x - 7$$

$$2x - 3x = -7 + 4$$

$$-x = -3$$

$$x = 3$$

Check: $$\frac{2x-4}{x-3} = 3 + \frac{2}{x-3}$$

$$\frac{6-4}{3-3} \overset{?}{=} 3 + \frac{2}{3-3}$$

But division by zero is undefined, so 3 cannot be a solution.

Answer: The equation has no solution.

_____

It's your turn again.

Problem 2

Solve for x: $\dfrac{3}{2x-1} = 5$.

_____

Solution:

_____

Fractional equations of the following type occur quite often:

$$\frac{2}{x+1} = \frac{3}{x}$$

By this I mean an equation with **two** fractions, one on each side of the equal sign, and no other terms. Such an equation is called a **proportion**. There is an easy way to solve this kind of fractional equation, that is, to "cross-multiply."

$$\frac{2}{x+1} \times \frac{3}{x}$$
$$2x = 3(x+1)$$

What we actually did was to multiply the entire equation by the lowest common denominator, $x(x+1)$. But by simply "cross-multiplying" we save ourselves a few steps. (If you don't believe me, try it the long way.) Then the problem continues on as before.

$$2x = 3x + 3$$
$$2x - 3x = 3$$
$$-x = 3$$
$$x = -3$$

Check:
$$\frac{2}{x+1} = \frac{3}{x}$$
$$\frac{2}{(-3)+1} \overset{?}{=} \frac{3}{(-3)}$$
$$\frac{2}{-2} \overset{?}{=} \frac{3}{-3}$$
$$-1 = -1$$

Answer: This is true. So $x = -3$ is the solution.

_____

EXAMPLE 5

Solve for x: $\dfrac{5}{2x-1} = \dfrac{3}{x+1}$.

Solution:

$$\dfrac{5}{2x-1} \diagdown \diagup \dfrac{3}{x+1}$$

$$5(x+1) = 3(2x-1)$$
$$5x+5 = 6x-3$$
$$5x-6x = -3-5$$
$$-x = -8$$
$$x = 8$$

Check:

$$\dfrac{5}{2x-1} = \dfrac{3}{x+1}$$

$$\dfrac{5}{2(8)-1} \overset{?}{=} \dfrac{3}{(8)+1}$$

$$\dfrac{5}{15} \overset{?}{=} \dfrac{3}{9}$$

$$\dfrac{1}{3} = \dfrac{1}{3}$$

Answer: This is true. So $x = 8$ is the solution.

EXAMPLE 6

Solve for x: $\dfrac{3x}{2x-3} = 4$.

Solution: Since a whole number can always be divided by 1 without changing its value, we can write 4 as $\dfrac{4}{1}$ and then cross-multiply.

$$\dfrac{3x}{2x-3} = \dfrac{4}{1}$$
$$3x(1) = 4(2x-3)$$
$$3x = 8x-12$$
$$12 = 8x-3x$$
$$12 = 5x$$
$$\dfrac{12}{5} = \dfrac{\cancel{5}x}{\cancel{5}}$$
$$x = \dfrac{12}{5} = 2.4$$

Check: $$\frac{3(2.4)}{2(2.4) - 3} = 4$$

$$4 = 4$$

(Use a calculator.)

Answer: $x = \dfrac{12}{5} = 2.4$

You should now be able to identify and solve fractional equations. Remember that an equation of this type has a variable in a denominator. In fact, there may be a fraction on either side or both sides of the equal sign.

The same four basic steps—simplify, transpose, simplify, and divide—are used to solve fractional equations. In addition, we must check the tentative solution by substituting it in the original equation.

Remember also that, if we make a wrong choice for the common denominator and a denominator remains after completing the basic steps, we can simply repeat the process, using the remaining denominator.

Finally, remember that, when there is a single fraction on both sides of the equal sign, we can use a shortcut: cross-multiplying the fractions.

Before beginning the next unit you should solve the following equations.

EXERCISES

Solve for x:

1. $1 = \dfrac{5}{x}$

2. $\dfrac{x-3}{2} = \dfrac{2x+4}{5}$

3. $\dfrac{6}{x-2} = -3$

4. $\dfrac{3x-3}{x-1} = 2$

5. $\dfrac{x}{2} = \dfrac{x+6}{3}$

6. $\dfrac{3}{x} = \dfrac{4}{x-2}$

7. $\dfrac{4}{x+3} = \dfrac{1}{x-3}$

8. $\dfrac{5-2x}{x-1} = -2$

9. $\dfrac{x+3}{x-2} = 2$

10. $\dfrac{4}{5}x - \dfrac{1}{4} = -\dfrac{3}{2}x$

11. $5 + \dfrac{3+x}{x} = \dfrac{5}{x}$

12. $\dfrac{4}{x-2} - \dfrac{1}{x} = \dfrac{5}{x-2}$

Hint: The common denominator is $x(x-2)$.

Solutions to Problems

1.
$$\frac{2}{x+1} + 3 = 1$$

$$(x+1)\left(\frac{2}{x+1} + 3 = 1\right)$$

$$\frac{\cancel{(x+1)}2}{\cancel{x+1}} + 3(x+1) = (x+1)$$

$$2 + 3(x+1) = x+1$$

$$2 + 3x + 3 = x+1$$

$$3x + 5 = x+1$$

$$3x - x = 1 - 5$$

$$2x = -4$$

$$\frac{\cancel{2}x}{\cancel{2}} = \frac{\cancel{-4}^{-2}}{\cancel{2}}$$

$$x = -2$$

Check:
$$\frac{2}{-2+1} + 3 = 1$$

$$-2 + 3 = 1$$

$$1 = 1$$

The solution is $x = -2$.

2.
$$\frac{3}{2x-1} = 5$$

$$(2x-1)\left(\frac{3}{2x-1} = 5\right)$$

$$\frac{\cancel{(2x-1)}3}{\cancel{2x-1}} = 5(2x-1)$$

$$3 = 5(2x-1)$$

$$3 = 10x - 5$$

$$3 + 5 = 10x$$

$$8 = 10x$$

$$\frac{\cancel{8}^4}{\cancel{10}_5} = \frac{\cancel{10}x}{\cancel{10}}$$

$$\frac{4}{5} = x$$

Check:
$$\frac{3}{2\left(\frac{4}{5}\right) - 1} = 5$$

$$5 = 5$$

(Use a calculator.)

The solution is $x = \frac{4}{5}$.

UNIT 5

Another Special Type of Equation: The Literal Equation

In this unit you will learn about another special type of equation, the literal equation. When you have completed the unit, you will be able to identify and solve literal equations.

Definition: A **literal equation** is an equation that contains letters and numbers. **Formulas** often are written as literal equations.

We solve a literal equation for the stated letter in terms of the other letters. That is, our answer will no longer be a simple numerical value but will contain some, or all, of the other letters in the equation.

Again, the basic strategy we have been using to solve other first-degree equations can be used to solve literal equations. I have expanded the steps slightly because there is more than one letter in the equation.

To solve a literal equation for a given letter, identify the letter and

1. **Simplify:**   clear any fractions,
   remove parentheses,
   collect like terms.

2. **Transpose:**
   Isolate all terms with the letter to be solved for on one side of the equation and transpose everything else to the other side.

3. **Simplify.**

4. **Divide** each term of the equation **by the coefficient** of the letter to be solved for.

This is an example of a literal equation:

$$2x - 4p = 3x + 2p$$

EXAMPLE 1

Solve this literal equation for x: $2x - 4p = 3x + 2p$.

Solution: $2x - 4p = 3x + 2p$

$$2x - 3x = 2p + 4p$$

$$-x = 6p$$

$$\frac{-x}{-1} = \frac{6p}{-1}$$

$$x = -6p$$

Now *you* solve the same equation for p.

Problem 1

Solve this literal equation for p: $2x - 4p = 3x + 2p$.

Solution:

Now we briefly turn our attention to some examples with formulas. The area of a triangle is given by the formula: $A = \frac{1}{2}bh$, where A is the area, b is the base, and h is the height.

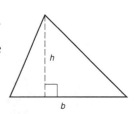

EXAMPLE 2

Solve this formula for b: $A = \frac{1}{2}bh$.

Solution:
$$A = \frac{1}{2}bh$$

$$2\left(A = \frac{1}{2}bh\right)$$

$$2 \cdot A = 2 \cdot \frac{1}{2}bh$$

$$2A = bh$$

$$\frac{2A}{h} = \frac{bh}{h}$$

$$\frac{2A}{h} = b$$

Problem 2

The formula for the perimeter of a rectangle is $P = 2l + 2w$, where P is the perimeter, l is the length, and w is the width.

Solve for w.

w

l

Solution:

Let's do two more examples.

EXAMPLE 3

Solve this equation for x: $ax - 3 = -2cx$

Solution:
$$ax - 3 = -2cx$$

$$ax - 3 = -2cx$$

Transpose: $ax + 2cx = 3$

Recall from Unit 2 the distributive property, $a(b + c) = ab + ac$, which we used to remove parentheses. Now, in order to find the coefficient of x, we will use the distributive property in reverse.

Notice that x is common to both terms on the left side of the equation. Using the distributive property, we can factor out the common x since

$$ax + 2cx = x(a + 2c)$$

Factor out x: $\quad x(a + 2c) = 3$

$$\frac{x(\cancel{a + 2c})}{\cancel{a + 2c}} = \frac{3}{a + 2c}$$

$$x = \frac{3}{a + 2c}$$

Notice that for our literal equations we do *not* get a nice simple number for an answer; instead we get a fairly complicated-looking expression for x in terms of the other letters.

EXAMPLE 4

Solve this equation for x: $\quad 3x + 5y = ax + 2y.$

Solution:

$$3x + 5y = ax + 2y$$
$$3x - ax = 2y - 5y$$
$$3x - ax = -3y$$

Factor out x.
$$x(3 - a) = -3y$$

$$\frac{x(\cancel{3 - a})}{\cancel{3 - a}} = \frac{-3y}{3 - a}$$

$$x = \frac{-3y}{3 - a}$$

Or, if you prefer, multiply top and bottom by -1.
$$x = \frac{3y}{a - 3}$$

More often than not we solve equations for y in terms of x. Using the same equation as in Example 4, try it.

Problem 3

Solve this equation for y: $\quad 3x + 5y = ax + 2y.$

Solution:

We'll do two more examples.

EXAMPLE 5

Solve this equation for y: $\dfrac{3y+a}{a} = \dfrac{4y+b}{b}$.

Solution:

$$\frac{3y+a}{a} = \frac{4y+b}{b}$$

Shortcut: cross-multiply from Unit 4. $b(3y + a) = a(4y + b)$

$$3by + ab = 4ay + ab$$

$$3by - 4ay = ab - ab$$

Factor out y. $y(3b - 4a) = 0$

Divide by the coefficient of y. $\dfrac{y(\cancel{3b - 4a})}{(\cancel{3b - 4a})} = \dfrac{0}{3b - 4a}$

$$y = 0$$

QUESTION: Must we now check to see whether this answer is indeed the solution to the equation? In other words, is this a fractional equation?

ANSWER: No, y is not in any of the denominators.

EXAMPLE 6

Solve for x: $a(x-1) = -\dfrac{x}{b}$.

Solution:

$$a(x - 1) = -\frac{x}{b}$$

$$b\left(a(x - 1) = \frac{-x}{b}\right)$$

$$ab(x - 1) = \frac{-x\cancel{b}}{\cancel{b}}$$

$$ab(x - 1) = -x$$

$$abx - ab = -x$$

$$abx + x = ab$$

Factor out x: $x(ab + 1) = ab$

$$\frac{x(\cancel{ab + 1})}{(\cancel{ab + 1})} = \frac{ab}{ab + 1}$$

$$x = \frac{ab}{ab + 1}$$

Now it's your turn to solve the above equation for b.

Problem 4

Solve for b: $a(x-1) = -\dfrac{x}{b}$.

Solution:

You should now be able to identify and solve first-degree literal equations for a particular letter. Remember that a literal equation is an equation that contains numbers and more than one letter. Formulas are often written as literal equations. Even though literal equations are a special type of equation, the same basic strategy is used to solve them.

Remember also that, as we learned in Unit 4, when there is a **single** fraction on both sides of a literal equation, we can reduce our work by cross-multiplying.

Before beginning the next unit you should solve the following literal equations.

EXERCISES

Solve for a:

1. $\dfrac{2ax}{3c} = \dfrac{y}{m}$
2. $2cy + 4d = 3ax - 4b$
3. $ax + 3a = bx + 7c$
4. $4x + 5c - 2a = 0$
5. $a(x+2) = \pi - cy$

6–10. Now solve each of the above equations for x.

11–15. Now solve each of the above equations for c.

16. $C = 2\pi r$ is the formula for the circumference of a circle. Solve for r.

17. $R = \dfrac{V}{I}$ is Ohm's law in electrical theory. Solve for I.

18. $A = \dfrac{1}{2}(a+b)h$ is the formula for the area of a trapezoid. Solve for a.

Solutions to Problems

1. $2x - 4p = 3x + 2p$

$2x - 3x = 2p + 4p$

$-x = 6p$

$\dfrac{-x}{6} = \dfrac{\cancel{6}p}{\cancel{6}}$

$\dfrac{-x}{6} = p$

2. $P = 2l + 2w$

$P - 2l = 2w$

$\dfrac{P - 2l}{2} = \dfrac{\cancel{2}w}{\cancel{2}}$

$\dfrac{P - 2l}{2} = w$

3. $3x + 5y = ax + 2y$

$5y - 2y = ax - 3x$

$3y = ax - 3x$

$\dfrac{\cancel{3}y}{\cancel{3}} = \dfrac{ax - 3x}{3}$

$y = \dfrac{ax - 3x}{3}$

4. $a(x - 1) = -\dfrac{x}{b}$

$b\left(a(x - 1) = \dfrac{-x}{b} \right)$

$ab(x - 1) = \dfrac{-x\cancel{b}}{\cancel{b}}$

$ab(x - 1) = -x$

$\dfrac{ab\cancel{(x-1)}}{\cancel{a}\cancel{(x-1)}} = \dfrac{-x}{a(x - 1)}$

$b = \dfrac{-x}{a(x - 1)}$

UNIT 6

Applied Problems

In this unit we will consider a few applied problems, sometimes referred to as word problems. First, you will learn how to translate a word problem into an algebraic equation, which then can be solved using the same basic strategy learned in previous units. The emphasis will be not only on the numerical answer but also on the interpretation of the answer in terms of the problem.

If you're tempted to skip this unit, don't. The problems really aren't that hard, and it is important to understand how algebra can be used to solve problems in the real world. We like to think of each one as a puzzle, and we enjoy solving puzzles.

CHANGING VERBAL STATEMENTS TO ALGEBRAIC EXPRESSIONS

In algebra, as you know, letters are used to represent numbers. By using letters and mathematical symbols, we can replace lengthy verbal statements with short algebraic expressions.

Here are a few examples.

EXAMPLE 1	The sum of a number and 7	$x + 7$
EXAMPLE 2	Five minus some number	$5 - y$
EXAMPLE 3	Seven times a number plus 3	$7x + 3$

EXAMPLE 4   A number divided by 11 $\dfrac{z}{11}$

EXAMPLE 5   Four less than a number $x - 4$

EXAMPLE 6   Eight more than a number $x + 8$

EXAMPLE 7   Half of a number $\dfrac{1}{2}x$ or $\dfrac{x}{2}$

You try the next few.

Let x represent a number. Write each of the following in terms of x.

Problem 1   Three times a number minus 2

Problem 2   Seven divided by a number

Problem 3   Five plus a number

Problem 4   Five less than twice a number

Problem 5   Ten more than half a number

CHANGING VERBAL STATEMENTS TO ALGEBRAIC EQUATIONS

We'll continue translating words into symbols, but now we'll include an equal sign.

EXAMPLE 8   Three times a number minus 2 equals 10. Find the number.

Solution: $3x - 2 = 10$

$3x = 10 + 2$

$3x = 12$

$\dfrac{\cancel{3}x}{\cancel{3}} = \dfrac{\cancel{12}^{4}}{\cancel{3}}$

$x = 4$

EXAMPLE 9   Five times a number plus 2 is 10. Find the number.

Solution: $5x + 2 = 10$ Notice that the word *is* translates to an equal sign.

$$5x = 10 - 2$$

$$5x = 8$$

$$\frac{5x}{5} = \frac{8}{5}$$

$$x = \frac{8}{5}$$

Here are three for you to try. For each problem, write an equation and then solve it.

Problem 6   Five plus a number is 7. Find the number.

Problem 7   When a number is decreased by 3, the result is 15. Find the number.

Problem 8   The sum of a number and 11 is 12. Find the number.

We think we're ready to look at some word problems.

SOLVING WORD PROBLEMS

When attempting to solve a word problem, we must translate the problem into algebraic symbols, write and solve the equation, and finally answer the original question. We'll demonstrate how to solve one or two word problems before we give you a similar problem to solve.

EXAMPLE 10

Bob is 5 years older than Barbara. The sum of their ages is 23. How old is Barbara?

Solution: Let x represent Barbara's age.

Then Bob's age is $x + 5$ because he is 5 years older than Barbara.

The sum of their ages is 23.

Bob's age + Barbara's age = 23

$$(x + 5) + \qquad\qquad x = 23$$
$$2x + 5 = 23$$
$$2x = 18$$
$$x = 9$$

Answer: Barbara is 9 years old.

Maybe you wanted to start the previous example by letting x represent Bob's age. Would this be wrong? No, because applied problems can often be approached from different directions, all of which lead to the correct solution. The next example should help convince you of that fact.

EXAMPLE 11

Resolve Example 10 letting x represent Bob's age.

Solution: Let x represent Bob's age.

Then Barbara's age is $x - 5$ because she is 5 years younger than Bob.

The sum of their ages is 23.

Bob's age + Barbara's age = 23

$$x \quad + \qquad (x - 5) = 23$$
$$2x - 5 = 23$$
$$2x = 28$$
$$x = 14$$

But the problem asked for Barbara's age.

Barbara's age is given by $x - 5$, thus $14 - 5 = 9$.

Answer: Barbara is 9 years old.

Problem 9

Laurie and Lynda are sisters. Lynda is 16 years older than Laurie. The sum of their ages is 60. How old is each sister?

Before continuing we need to introduce some new terminology with respect to integers. Recall that the set of integers is $\{\ldots, -4, -3, -2, -1, 0, 1, 2, 3, 4, \ldots\}$.

Examples of **consecutive integers** are 21, 22, 23, 24 and −5, −4, −3. Notice that if x represents an integer, a set of consecutive integers can be represented as $x, x+1, x+2, x+3$, and so on.

Examples of **consecutive even integers** are 10, 12, 14, 16 and −4, −2, 0, 2, 4. Notice that if x represents an even integer, a set of even consecutive integers can be represented as $x, x+2, x+4$, and so on.

Examples of **consecutive odd integers** are 37, 39, 41 and −7, −5, −3, −1, 1, 3. Notice that if x represents an odd integer, a set of consecutive odd integers can be represened as $x, x+2, x+4$, and so on. Consecutive odd integers, like consecutive even integers, always differ by 2.

EXAMPLE 12

On U.S. Route 1 in the Florida Keys, the mile marker numbers increase from the south (Key West) to the north (Key Largo). The sum of two consecutive mile markers on U.S. Route 1 in the Florida Keys is 115. Find the numbers on the markers.

Solution: Let x represent the smaller integer.

Then the next consecutive integer is $x + 1$.

The sum of the two consecutive mile markers is 115.

$$x + (x+1) = 115$$
$$2x + 1 = 115$$
$$2x = 114$$
$$x = 57$$
$$\text{and } x + 1 = 58$$

Answer: The mile marker numbers are 57 and 58.

EXAMPLE 13

Jerry and Caroline Cash ordered tickets for adjacent seats for the Paradise Big Band's final concert of the season. The San Carlos Institute, a small auditorium where concerts are held, has one center aisle. Seats on the right of the aisle are numbered by consecutive even numbers, and on the left by consecutive odd numbers. When the tickets arrive, Jerry notices that the sum of the two seat numbers is 208. Are the Cashes' seats on the right or left of the center aisle and what are their seat numbers?

_____

Solution: Let x represent the first seat number.

Then the next seat number is $x + 2$ because it is either a consecutive even or consecutive odd number.

The sum of the two seat numbers is 208.

$$x + (x + 2) = 208$$
$$2x + 2 = 208$$
$$2x = 206$$
$$x = 103$$
$$x + 2 = 105$$

Answer: Jerry and Caroline Cash will be seated on the left side of the aisle in seats 103 and 105.

_____

I have done two short examples; now here's a longer one for you.

Problem 10

Rob lives in Pittsburgh. His baseball team is raffling off a ten-speed bike to raise money to update the team's mascot. Rob bought five raffle tickets, which are numbered consecutively. The sum of the raffle ticket numbers is 1075. What are the numbers on his raffle tickets?

_____

Solution:

Many application problems involve formulas we already know. A familiar formula that relates the concepts of distance, speed, and time is $rt = d$. For example, if you drive on the turnpike for 2 hours at an average speed of 60 mph, you will have traveled a distance of 120 miles, or 2 times 60 = 120. We will use this formula in the next two examples, which are often referred to as motion problems.

EXAMPLE 14

Patti and Tom decide to go for a walk. Patti likes to walk fast, and Tom likes to walk slowly. Patti starts out walking due north at 3 mph, while Tom decides to head due south, ambling along at 1 mph. After how many hours will they be 5 miles apart?

Solution: We find it helpful to construct a chart for motion problems of this type. The formula is shown across the top, and the given information about the rates has been filled in.

	rate	×	time	= distance
Patti	3 mph			
Tom	1 mph			

Let t represent Patti's walking time in hours.

Then Tom's walking time also will be t because they are starting and stopping at the same time.

The final column of the chart is found by multiplying across each row, that is, rate × time = distance.

	rate	×	time	= distance
Patti	3 mph		t	$3t$
Tom	1 mph		t	t

Now we are ready to try writing the equation needed to solve the problem. Typically the required information comes from the two entries in the distance column. You need to ask yourself a few questions. Will the distances walked by Patti and Tom be equal? No, because they are walking at different rates but for the same amount of time. Do we know what the sum of the distances walked will equal? Yes, the sum of the distances walked will be 5 miles because we want to know when they will be 5 miles apart.

$$\text{Patti's distance} + \text{Tom's distance} = 5$$

$$3t + t = 5$$

$$4t = 5$$

$$t = \frac{5}{4}$$

$$t = 1.25$$

Answer: After walking for 1.25 hours (or 1 hour and 15 minutes), Patti and Tom will be 5 miles apart.

_____

The final example, while still using $d = rt$, has a few twists to it, so be careful.

EXAMPLE 15

Orvis Kemp owns a 1968 red Mercedes roadster in mint condition. Orvis leaves his home in Key West and drives at a constant rate of 40 mph. One hour later, Rita, his wife, leaves in her new Mercedes and travels on the same route at a constant rate of 50 mph. How long will it take for Rita to catch up with Orvis?

_____

Solution: We're going to work this one together. Without looking at the completed chart, try filling in the blank chart on your own.

	rate	×	time	= distance
Orvis Rita				

	rate	×	time	= distance
Orvis	40 mph		t	$40t$
Rita	50 mph		$t - 1$	$50(t - 1)$

Did you realize that in this example the times are not equal? Rita starts out 1 hour later; thus she will be driving 1 hour less than Orvis. Now for the equation needed to solve the problem. What is true about the distances driven by each person? In this problem the distances are equal.

distance driven by Orvis = distance driven by Rita

$$40t = 50(t - 1)$$
$$40t = 50t - 50$$
$$50 = 50t - 40t$$
$$50 = 10t$$
$$\frac{50}{10} = \frac{10t}{10}$$
$$5 = t$$

Answer: Rita will catch up with Orvis in 4 hours. Note that the answer is 4, not 5, because the question asked about Rita's driving time, which was $t - 1$.

———

We think we've done enough examples.

By now you should be able to translate verbal expressions into algebraic symbols, and to translate an applied problem into an equation.

Before going on to the next unit, do the following exercises.

EXERCISES

Let x represent a number. Express each of the following in terms of x:

1.   A number decreased by 5.

2.   Three times a number increased by 8.

3.   Eight times a number minus 10.

4.   A number divided by 3.

Let x represent a number. Express each of the following as an equation:

5.   Two times a number decreased by 5 equals 11.

6.   Seven times a number is 35.

7.   Twenty less than a number is 32.

8.   The sum of x and 12 is 20.

9.   Fifteen increased by 2 times a number is 47.

10.   Four more than three times a number is 17.

Solve each of the following:

11. George is 8 years older than Jack. The sum of their ages is 42. How old is each person?

12. A rope that is 36 feet in length is cut into two pieces. If one piece is 10 feet longer than the other, how long is each piece?

13. The Cancer Foundation of Denver is raffling off a car. The raffle tickets are $100 each. Mercy Hiller and her husband have bought four tickets, which they notice are numbered consecutively. The sum of the numbers is 1354. What are the numbers on their raffle tickets?

14. The sum of three consecutive even integers is 6480. What are the integers?

15. The formula for the perimeter of a rectangle is $P = 2l + 2w$. The perimeter of the local high school basketball court is 268 ft. The length is 34 ft longer than the width. Find the dimensions of the basketball court.

16. The perimeter of the Searstown parking lot is 4100 ft. The length is 3 times the width. Find the dimensions of the parking lot.

17. Aurora and Chris work together at the local campus bookstore. They live 4 miles apart on the same road. Aurora starts walking at 3 mph, and at the same instant Chris starts walking toward her at 2 mph. When will they meet?

18. Every morning Rosalie runs along a runner's path on Pawley's Island at the constant rate of 6 mph. A half-hour later her friend, Sally, begins at the same point, running at a rate of 8 mph and following the same path. How long will it take Sally to catch up to Rosalie?

Solutions to Problems

1. $3x - 2$

2. $\dfrac{7}{x}$

3. $5 + x$

4. $2x - 5$

5. $\dfrac{1}{2}x + 10$

6. $5 + x = 7$

$x = 7 - 5$

$x = 2$

7. $x - 3 = 15$

$x = 15 + 3$

$x = 18$

8. $x + 11 = 12$

$x = 12 - 11$

$x = 1$

9. Let x represent Laurie's age.

Then Lynda's age is $x + 16$ because she is 16 years older than Laurie.

The sum of their ages is 60.

Laurie's age + Lynda's age = 60

$\quad x \qquad + \qquad x + 16 = 60$

$2x + 16 = 60$

$2x = 60 - 16$

$2x = 44$

$\dfrac{\cancel{2}x}{\cancel{2}} = \dfrac{\cancel{44}^{\,22}}{\cancel{2}}$

$x = 22$

Laurie is 22 years old and Lynda is $x + 16$ or $22 + 16 = 38$ years old.

10. Let x represent the first ticket number.

Then the next four ticket numbers are $x + 1$, $x + 2$, $x + 3$, and $x + 4$ because they are consecutive numbers.

The sum of the ticket numbers is 1075.

$x + (x + 1) + (x + 2) + (x + 3) + (x + 4) = 1075$

$5x + 10 = 1075$

$5x = 1075 - 10$

$5x = 1065$

$\dfrac{\cancel{5}x}{\cancel{5}} = \dfrac{\cancel{1065}^{\,213}}{\cancel{5}}$

$x = 213$

$x + 1 = 214$

$x + 2 = 215$

$x + 3 = 216$

$x + 4 = 217$

The ticket numbers are 213, 214, 215, 216, and 217.

UNIT 7

Positive Integral Exponents

In this unit you will learn to simplify expressions involving positive integral exponents. You will also learn four of the **five basic laws of exponents**. When you have completed the unit, you will be able to simplify expressions in which there is an exponent that is a positive integer or 0.

RECOGNIZING EXPONENTS

Consider the expression b^n. This is read as "b to the nth power." The b is referred to as the **base**, and the n is the **exponent**. Here $b \neq 0$.

In this unit we will consider only exponents that are 0 or a positive integer (1, 2, 3, 4, etc.). Negative and fractional exponents will be discussed in later units.

> Definition: **Positive Integral Exponent**
>
> $b^n = \underbrace{b \cdot b \cdot b \cdot \ldots \cdot b}_{n \text{ factors}}$, if n is a positive integer

Consider the positive integral exponents in these expressions.

EXAMPLE 1 $b^2 = b \cdot b$

EXAMPLE 2 $x^3 = x \cdot x \cdot x$

EXAMPLE 3 $2^5 = 2 \cdot 2 \cdot 2 \cdot 2 \cdot 2 = 32$

Any nonzero quantity raised to the 0 power is 1.

Definition: **Zero Exponent**
$$b^0 = 1, \text{ where } b \neq 0$$

EXAMPLE 4 $x^0 = 1, x \neq 0,$

EXAMPLE 5 $5^0 = 1$

EXAMPLE 6 $\left(3ab + \pi - 5\sqrt{7}\right)^0 = 1$

ORDER OF OPERATIONS, AGAIN

Now that we have defined exponents, we need to revise our order of operations to include them. Evaluating exponents comes after simplifying inside parentheses and before multiplication and division.

Step 1:   First perform any operations inside parentheses. If there is more than one operation inside the parentheses, do them according to the order of operations.

Step 2:   Evaluate or simplify exponents.

Step 3:   Perform multiplications and divisions from left to right.

Step 4:   Finally, perform additions and subtractions from left to right.

EXAMPLE 7

Evaluate: $6 + 4 \cdot 3^2$

Solution: There are no parentheses, so start with Step 2.

$$
\begin{aligned}
6 + 4 \cdot 3^2 &= 6 + 4 \cdot 9 \quad &\text{Exponent} \\
&= 6 + 36 \quad &\text{Multiplication} \\
&= 42 \quad &\text{Addition}
\end{aligned}
$$

EXAMPLE 8

Evaluate: $(14 - 10 \div 2)^3$

Solution: $(14 - 10 \div 2)^3 = (14 - 5)^3$ Inside parentheses, division
$= 9^3$ Inside parentheses, subtraction
$= 729$ Exponent

Try one yourself.

Problem 1 $3(6 - 2)^3$

SIMPLIFYING EXPRESSIONS WITH EXPONENTS

To simplify, we have five basic laws of exponents. We will discuss only four at this time.

These laws are used to shorten our work. When in doubt about any of these laws, we can always go back to the definitions on the first and second pages of this unit.

Laws of Exponents

I. **Multiplication** $b^n \cdot b^m = b^{n+m}$

II. **Power of a power** $(b^n)^m = b^{nm}$

III. **Power of a product** $(bc)^n = b^n c^n$

IV. **Power of a fraction** $\left(\dfrac{a}{b}\right)^n = \dfrac{a^n}{b^n}$

Let's do examples for each of these laws.

I. **Multiplication** $b^n \cdot b^m = b^{n+m}$

EXAMPLE 9

Simplify: $a^2 \cdot a^3$.

Solution: $a^2 \cdot a^3 = a^{2+3} = a^5$

If you doubt this, use the definition: $a^2 \cdot a^3 = (a \cdot a) \cdot (a \cdot a \cdot a) = a^5$.

Note that if no exponent is written, the exponent is understood to be 1.

EXAMPLE 10

Simplify: $x \cdot x^2$.

Solution: $x \cdot x^2 = x^1 \cdot x^2 = x^{1+2} = x^3$

EXAMPLE 11

Simplify: $5 \cdot 5^2$.

Solution: $5 \cdot 5^2 = 5^1 \cdot 5^2 = 5^{1+2} = 5^3$

In case you're not convinced:

$5 \cdot 5^2 = 5 \cdot (5 \cdot 5) = 5^3$ Note: the base does not change.

EXAMPLE 12

Simplify: $a^2(2a^3)$.

Solution: $a^2(2a^3) = a^2 \cdot 2 \cdot a^3 = 2a^5$ Note: 2 is the coefficient and not the base.

Before doing the next example, here is a reminder from Unit 2. When a term contains a number and several letters, the number is written first and the letters usually are put in alphabetical order.

EXAMPLE 13

Simplify: $(5a^2b^3)(ab^4)(3abc)$.

Solution: $(5a^2b^3)(ab^4)(3abc)$

$$5a^2 \cdot b^3 \cdot a^1 \cdot b^4 \cdot 3 \cdot a^1 \cdot b^1 \cdot c^1 = 5 \cdot 3a^2 \cdot a^1 \cdot a^1 \cdot b^3 \cdot b^4 \cdot b^1 \cdot c^1$$

$$= 15a^{2+1+1}b^{3+4+1}c^1$$

$$= 15a^4b^8c$$

You try some.

Problem 2 $\quad x^3 \cdot x^5 =$

Problem 3 $\quad x^2(x^3 y) =$

Problem 4 $\quad (x^2 y^3)(x^7 y) =$

Problem 5 $\quad (2w^3 a^5)(3a^2 w) =$

$$\boxed{\text{II. } \textbf{Power of a power} \quad (b^n)^m = b^{nm}}$$

====

EXAMPLE 14

Simplify: $(a^2)^3$.

Solution: $\quad (a^2)^3 = a^{2 \cdot 3} = a^6$

If we had used the definition instead:

$$(a^2)^3 = (a^2)(a^2)(a^2)$$
$$= a^{2+2+2}$$
$$= a^6$$

Note that the laws of exponents are shortcut methods. With them, you do not have to work out expressions completely with the definition. But they are shortcuts only if you apply them properly.

====

EXAMPLE 15

Simplify: $(x^{15})^2$.

Solution: $\quad (x^{15})^2 = x^{15 \cdot 2} = x^{30}$

Now it's your turn again.

Problem 6 $\quad (a^{10})^2 =$

Problem 7 $\quad (x^3)^0 =$

Problem 8 $\quad c^{10} \cdot c^2 =$

Problem 9 $\quad (w^2)^4 =$

> III. **Power of a product** $(bc)^n = b^n c^n$

EXAMPLE 16

Simplify: $(3x^2)^2$.

Solution: $(3x^2)^2 = 3^2 x^4 = 9x^4$

Be careful; most people forget that the 3 must also be squared.

EXAMPLE 17

Simplify: $(2x^2 y^3)^4$.

Solution: $(2x^2 y^3)^4 = 2^4 x^{2 \cdot 4} y^{3 \cdot 4} = 2^4 x^8 y^{12} = 16 x^8 y^{12}$

> IV. **Power of a fraction** $\left(\dfrac{a}{b}\right)^n = \dfrac{a^n}{b^n}$

EXAMPLE 18

Simplify: $\left(\dfrac{x^2}{c^3}\right)^5$.

Solution: $\left(\dfrac{x^2}{c^3}\right)^5 = \dfrac{x^{2 \cdot 5}}{c^{3 \cdot 5}} = \dfrac{x^{10}}{c^{15}}$

EXAMPLE 19

Simplify: $\left(\dfrac{7w^3}{5s}\right)^2$.

Solution: $\left(\dfrac{7w^3}{5s}\right)^2 = \dfrac{7^2 w^{3 \cdot 2}}{5^2 s^2} = \dfrac{49w^6}{25s^2}$

Have you noticed that the power laws deal with removing parentheses? To remove the parentheses we multiply the exponents.

Sound familiar? To remove parentheses, we multiply. . . .

As the algebraic expressions become more complex, and you know they will, it often becomes necessary to use several laws in the same problem. My choice of procedure **always** will be to remove the parentheses first (power laws) and finish simplifying the expression by using the multiplication law.

EXAMPLE 20

Simplify: $a^2(4a^3)^2$.

Solution: $a^2(4a^3)^2 = a^2 \cdot 4^2 a^{3 \cdot 2} = 16a^2 a^6 = 16 \cdot a^{2+6} = 16a^8$

EXAMPLE 21

Simplify: $x^3(3\pi x)^0$.

Solution: $x^3(3\pi x)^0 = x^3 \cdot 1 = x^3$

Try another yourself.

Problem 10

Simplify: $x^3(xy^2)^2$.

Solution:

ALGEBRA AND THE CALCULATOR (Optional)

Evaluating expressions with exponents on the graphing calculator requires careful use of parentheses. Omitting necessary parentheses will give the wrong answer.

EXAMPLE 22

Evaluate: x^2 for $x = -3$.

_____

Solution: Mentally, you can calculate the answer to be +9. If you omit the parentheses on the calculator, you will get the wrong answer, –9. The calculator is performing the correct order of operations for -3^2. The exponent operation, 3^2, is done first; this result is then multiplied by –1. Type –3 in parentheses for the correct evaluation.

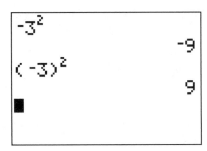

_____

You should now be able to simplify any expression in which an exponent of 0 or some positive integer appears. Remember that, when you are in doubt about one of the laws of exponents, you can always return to the definition of a positive integral exponent or a zero exponent.

Now try to simplify the following expressions. Be sure to check your answers with those at the back of the book and correct any mistakes before continuing on to the next unit.

EXERCISES

Simplify:

1. $(3y)^2$

2. $3x^0$

3. $x^2(x^3)^4$

4. $(x^2y^3z)^4$

5. $\left(\dfrac{x^2}{wz}\right)^3$

6. $(2ab)b^2$

7. $\dfrac{\left(x^2y^3\right)^2}{5}$

8. $5(x^2z)^2$

9. $(5x^2z)^2$ Be sure to notice that Problems 8 and 9 are different.

10. $\left(\dfrac{a^2b^3cd^5}{3x^2w^0}\right)^7$

11. $\dfrac{(2ab)^2}{\left(3x^3\right)^2}$

12. $(3x^5)^2(2x^3)^3$

13. $(x^2y)(xy^2)$

14. $2(3ab^2)^2$

15. $(-4c)^2$

16. $\left(\dfrac{xyz^2}{5a}\right)^3$

17. $(-2abc)(bcd)(3abc^2)$

18. $(2x^2yz)(-5xz)^2(xyz^2)^3$

- -

Solutions to Problems

1. $3(6-2)^3 = 3(4)^3 = 3 \cdot 64 = 192$

2. $x^3 \cdot x^5 = x^{3+5} = x^8$

3. $x^2(x^3y) = x^{2+3}y = x^5y$

4. $(x^2y^3)(x^7y) = (x^2y^3)(x^7y^1)$
   $\qquad\qquad = x^{2+7}y^{3+1}$
   $\qquad\qquad = x^9y^4$

5. $(2w^3a^5)(3a^2w) = (2w^3a^5)(3a^2w^1)$
   $\qquad\qquad\qquad = 2 \cdot 3a^{5+2}w^{3+1}$
   $\qquad\qquad\qquad = 6a^7w^4$

6. $(a^{10})^2 = a^{10 \cdot 2} = a^{20}$

7. $(x^3)^0 = x^{3 \cdot 0} = x^0 = 1$

8. $c^{10} \cdot c^2 = c^{10+2} = c^{12}$

9. $(w^2)^4 = w^{2 \cdot 4} = w^8$

10. $x^3(xy^2)^2 = x^3x^2y^{2 \cdot 2}$
    $\qquad\qquad = x^{3+2}y^4$
    $\qquad\qquad = x^5y^4$

- -

UNIT 8

Negative Exponents

In this unit you will learn how to work with **negative exponents**. When you have completed the unit, you will be able to simplify expressions with negative exponents.

Definition: **Negative exponent**

$$b^{-n} = \frac{1}{b^n} \quad \text{and} \quad \frac{1}{b^{-n}} = b^n$$

where $b \neq 0$

The right-hand part of the above statement follows from the fact that

$$\frac{1}{b^{-n}} = \frac{1}{\frac{1}{b^n}} = 1 \div \frac{1}{b^n} = 1 \cdot b^n = b^n$$

Therefore a base with a negative exponent can be rewritten with a positive exponent by moving the base from the numerator to the denominator (or vice versa).

Consider Examples 1–12 and be certain you understand the simplification.

EXAMPLE 1 $2^{-1} = \dfrac{1}{2}$

EXAMPLE 2 $a^{-1} = \dfrac{1}{a}$

EXAMPLE 3 $b^{-2} = \dfrac{1}{b^2}$

EXAMPLE 4 $c^{-3} = \dfrac{1}{c^3}$

EXAMPLE 5 $2x^{-1} = \dfrac{2}{x}$ Note: $2x^{-1} \neq \dfrac{1}{2x}$

EXAMPLE 6 $3ab^{-2} = \dfrac{3a}{b^2}$

EXAMPLE 7 $5^{-1}ab^{-2}c = \dfrac{ac}{5b^2}$

EXAMPLE 10 $\dfrac{a^{-2}}{b^{-3}} = \dfrac{b^3}{a^2}$

EXAMPLE 8 $\dfrac{1}{a^{-2}} = a^2$

EXAMPLE 11 $\dfrac{y^{-3}}{x^2} = \dfrac{1}{y^3 x^2}$

EXAMPLE 9 $\dfrac{5}{x^{-3}y} = \dfrac{5x^3}{y}$

EXAMPLE 12 $\dfrac{-3x^{-2}z}{w} = \dfrac{-3z}{wx^2}$

A common mistake is to think that the −3 in Example 12 can be rewritten as a positive 3 by moving it to the denominator of the fraction. This would be wrong! The −3 is not an exponent; consequently the laws of exponents do not apply.

––

When simplifying expressions dealing with exponents, the objective is usually to write the final answer *without* zero or negative exponents. Luckily the laws of exponents introduced in Unit 7 apply for all types of exponents. We can use these laws and our definition to simplify expressions with negative exponents.

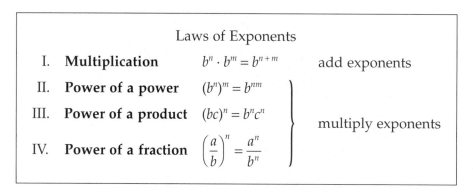

Laws of Exponents

I.   **Multiplication** $b^n \cdot b^m = b^{n+m}$ add exponents

II. **Power of a power** $(b^n)^m = b^{nm}$

III. **Power of a product** $(bc)^n = b^n c^n$ multiply exponents

IV. **Power of a fraction** $\left(\dfrac{a}{b}\right)^n = \dfrac{a^n}{b^n}$

With the introduction of negative exponents, algebraic expressions can become even more complex. Often there are several ways to simplify such complicated expressions. To make it easier for you to follow our solutions, we **always** will use this procedure when simplifying:

1. If there are parentheses, remove them using the power laws.

2. If there are negative exponents, rewrite them as positive exponents using the definition.

3. If necessary, finish by using the multiplication law of exponents.

   Before proceeding you might find it helpful to quickly review Unit 1 on adding, subtracting, dividing, and especially multiplying signed numbers.

═══

EXAMPLE 13

Simplify: $(10^{-3})^2$.

––

Solution: $(10^{-3})^2 = 10^{-6} = \dfrac{1}{10^6} = \dfrac{1}{1,000,000}$

––

EXAMPLE 14

Simplify: $(xy^{-1})^{-3}$.

Solution: $(xy^{-1})^{-3} = x^{-3}y^3 = \dfrac{y^3}{x^3}$

EXAMPLE 15

Simplify: $\dfrac{ab^{-4}}{a^{-2}b}$.

Solution: $\dfrac{ab^{-4}}{a^{-2}b} = \dfrac{a \cdot a^2}{b^4 \cdot b} = \dfrac{a^3}{b^5}$

EXAMPLE 16

Simplify: $\left(\dfrac{x^2}{y^{-3}w}\right)^{-1}$

Solution: $\left(\dfrac{x^2}{y^{-3}w}\right)^{-1} = \dfrac{x^{-2}}{y^3 w^{-1}} = \dfrac{w}{x^2 y^3}$

Now try four short problems on your own.

Problem 1

Simplify: $3a^{-2}x^5$.

Solution:

Problem 2

Simplify: $\dfrac{7x^{-3}wz^2}{x^4}$.

Solution:

Problem 3

Simplify: $\left(\dfrac{3xy^{-1}}{y}\right)^2$.

Solution:

Problem 4

Simplify: $\left(\dfrac{a}{b}\right)^{-1}$.

Solution:

Notice what happened—the fraction is simply inverted. *You have just proved a theorem!*

Hence

$$\left(\frac{a}{b}\right)^{-1} = \frac{b}{a}$$

We'll end the unit with three more examples.

EXAMPLE 17 $\left(\dfrac{3}{4}\right)^{-1} = \dfrac{4}{3}$

EXAMPLE 18 $\left(\dfrac{a^2b}{c^3}\right)^{-1} = \dfrac{c^3}{a^2b}$

Try the next one yourself before looking at the answer.

EXAMPLE 19

Simplify: $\left(\dfrac{x^2y^0}{w^{-1}}\right)^{-2}$.

Solution: $\left(\dfrac{x^2y^0}{w^{-1}}\right)^{-2} = \dfrac{x^{-4}y^0}{w^2} = \dfrac{1}{w^2x^4}$

You should now be able to simplify expressions involving negative exponents. Try simplifying the expressions in the exercises by writing them without negative exponents or zero exponents or parentheses.

EXERCISES

Simplify:

1. $\dfrac{a^{-3}}{a^2}$

2. $\dfrac{a^{-2}x^3}{y^{-1}}$

3. $(x^2y)^{-2}$

4. $x^5 \cdot x^0 \cdot z^{-7}$

5. $\dfrac{x^{-2}}{y^{-3}}$

6. $-2x^{-6}y^0$

7. $(3x^{-6}y^5)^{-2}$

8. $\dfrac{(ab^2)^{-3}}{(x^2y^{-3})^4}$

9. $\dfrac{(3ab^5)^{-3}}{2x^{-5}}$

10. $\dfrac{7x^{-1}}{y^2}$

11. $(5w^{-2})^2(2w^{-2})$

12. $\dfrac{x^{-2}y^{-3}}{c^{-2}}$

13. $\dfrac{(5a^2b^3)^2}{(-2x)^{-3}}$

14. $\dfrac{16w^{-1}y^2z^{-3}}{2x}$

15. $\left[\dfrac{b^2}{(a^2b)^{-2}}\right]^{-1}$

Solutions to Problems

1. $3a^{-2}x^5 = \dfrac{3x^5}{a^2}$

2. $\dfrac{7x^{-3}wz^2}{x^4} = \dfrac{7wz^2}{x^3x^4}$

   $= \dfrac{7wz^2}{x^{3+4}}$

   $= \dfrac{7wz^2}{x^7}$

3. $\left(\dfrac{3xy^{-1}}{y}\right)^2 = \dfrac{3^2x^2y^{-2}}{y^2}$

   $= \dfrac{3^2x^2}{y^2y^2}$

   $= \dfrac{9x^2}{y^{2+2}}$

   $= \dfrac{9x^2}{y^4}$

4. $\left(\dfrac{a}{b}\right)^{-1} = \dfrac{a^{-1}}{b^{-1}} = \dfrac{b}{a}$

UNIT 9

Division of Powers

In this unit you will learn to simplify expressions in which the same variables with exponents appear in both the numerator and the denominator. When you have completed the unit, you will be able to simplify expressions involving **division of like variables** raised to integral exponents.

Recall the four laws of exponents from Unit 7.

Laws of Exponents

I. **Multiplication** $\quad b^n \cdot b^m = b^{n+m}$ $\qquad$ add exponents

II. **Power of a power** $\quad (b^n)^m = b^{nm}$

III. **Power of a product** $\quad (bc)^n = b^n c^n$ $\qquad\Big\}$ multiply exponents

IV. **Power of a fraction** $\quad \left(\dfrac{a}{b}\right)^n = \dfrac{a^n}{b^n}$

We will now add a fifth and final law of exponents, which deals with division:

V. **Division:** $\quad \dfrac{b^m}{b^n} = b^{m-n}$

$\qquad\qquad$ or, alternatively, $\qquad\Big\}$ subtract exponents

$\qquad\qquad = \dfrac{1}{b^{n-m}}$

From a brief analysis of the laws we can produce a convenient way to classify them.

 I. Deals with multiplication—exponents added.
 II.
 III. } "In a sense" deal with removing parentheses—exponents multiplied.
 IV.
 V. Deals with division—exponents subtracted.

EXAMPLE 1

Simplify: $\dfrac{x^5}{x^3}$.

Solution: $\dfrac{x^5}{x^3} = x^{5-3} = x^2$

An alternative and longer solution uses the definition of positive integral exponents:

$$\frac{x^5}{x^3} = \frac{x \cdot x \cdot \cancel{x} \cdot \cancel{x} \cdot \cancel{x}}{\cancel{x} \cdot \cancel{x} \cdot \cancel{x}} = x \cdot x = x^2$$

EXAMPLE 2

Simplify: $\dfrac{a^4}{a^3}$.

Solution: $\dfrac{a^4}{a^3} = a^{4-3} = a^1 = a$

What about $\dfrac{x}{x^4}$? So far we have been using only the first half of the division law of exponents. Now consider a situation where the exponent in the denominator is larger than the exponent in the numerator.

Recall that, when simplifying an expression, we are attempting to write the final answer without using zero or negative exponents or parentheses.

EXAMPLE 3

Simplify: $\dfrac{x}{x^4}$.

Solution: $\dfrac{x}{x^4} = \dfrac{x^1}{x^4} = x^{1-4} = x^{-3} = \dfrac{1}{x^3}$

or, alternatively,

$$\frac{x}{x^4} = \frac{1}{x^{4-1}} = \frac{1}{x^3}$$

When the larger exponent is in the denominator, it takes one less step to use the second half of the division law of exponents—subtracting in the denominator.

EXAMPLE 4

Simplify: $\dfrac{a^2}{a^{10}}$.

Solution: $\dfrac{a^2}{a^{10}} = \dfrac{1}{a^{10-2}} = \dfrac{1}{a^8}$

Now consider these examples, which have more than one variable.

EXAMPLE 5

Simplify: $\dfrac{a^2 b^7 c^3}{a^5 b^2 c^4}$.

Solution: $\dfrac{a^2\,b^7\,c^3}{a^5\,b^2\,c^4} = \dfrac{b^{7-2}}{a^{5-2} c^{4-3}} = \dfrac{b^5}{a^3 c}$

EXAMPLE 6

Simplify: $\dfrac{a^2 b c^3}{a^7 b^3 c^3}$.

Solution: $\dfrac{a^2\,b\,c^3}{a^7\,b^3\,c^3} = \dfrac{1}{a^{7-2} b^{3-1} c^{3-3}} = \dfrac{1}{a^5 b^2}$

The c does not appear since $c^{3-3} = c^0 = 1$.

Again, let's move on to some problems involving definitions and the five laws of exponents.

To accomplish this, as in previous units, we **always** will use the same procedure when simplifying. Of course, an additional step has been added to include division.

Our suggested procedure for simplifying an algebraic expression with exponents is:

1. If there are parentheses, remove them using the power laws.

2. If there are negative exponents, use the definition to rewrite them as positive exponents.

3. If applicable, use the division law of exponents.

4. If applicable, finish by using the multiplication law of exponents.

There is nothing sacred about the order of these steps. In fact, they are completely interchangeable. You may do them in any order you wish. We suggest only that you *establish a pattern of your own and stick with it*.

EXAMPLE 7

Simplify: $\left(\dfrac{xy^0}{x^{-5}}\right)^{-2}$.

Solution: $\dfrac{x^{-2}y^0}{x^{10}} = \dfrac{x^{-2}}{x^{10}} = \dfrac{1}{x^2 x^{10}} = \dfrac{1}{x^{12}}$

EXAMPLE 8

Simplify: $\dfrac{\left(2^2\right)^{-1}}{\left(2^{-4}\right)^2}$.

Solution: $\dfrac{\left(2^2\right)^{-1}}{\left(2^{-4}\right)^2} = \dfrac{2^{-2}}{2^{-8}} = \dfrac{2^8}{2^2} = 2^{8-2} = 2^6 = 64$

EXAMPLE 9

Simplify: $\left(\dfrac{x^2 y}{xy^{-4}}\right)^3$.

Solution: $\left(\dfrac{x^2 y}{xy^{-4}}\right)^3 = \dfrac{x^6 y^3}{x^3 y^{-12}} = \dfrac{x^6 y^3 y^{12}}{x^3} = x^{6-3} y^{3+12} = x^3 y^{15}$

Now try two problems yourself.

Problem 1

Simplify: $\dfrac{x^9 y^{-6}}{x^{-3} y^2}$.

Solution:

Problem 2

Simplify: $\left(\dfrac{2x^{-1}y^2}{x^{-3}}\right)^2$.

Solution:

You should now be able to simplify expressions in which it is necessary to divide like variables with integral exponents. You also should be able to apply the definitions of integral exponents and to apply all five laws of exponents when simplifying expressions.

The classification scheme we developed for these laws, which are written to the right, should help you remember them.

Laws of Exponents

I.	**Multiplication**	$b^n \cdot b^m = b^{n+m}$	add exponents
II.	**Power of a power**	$(b^n)^m = b^{nm}$	
III.	**Power of a product**	$(bc)^n = b^n c^n$	multiply exponents
IV.	**Power of a fraction**	$\left(\dfrac{a}{b}\right)^n = \dfrac{a^n}{b^n}$	

V.	**Division**	$\dfrac{b^m}{b^n} = b^{m-n}$	
		or, alternatively,	subtract exponents
		$= \dfrac{1}{b^{n-m}}$	

Finally, we recommend establishing, and adhering to, your own pattern for simplification of expressions with exponents, such as ours. Before beginning the next unit you should simplify the expressions in the exercises. The final answers should be written without negative or zero exponents. The more of these problems you do, the easier they should become. When you have completed the exercises, check your answers against those given at the back of the book.

EXERCISES

Simplify:

1. $\dfrac{5a^7 b^2}{ab^{10}}$

2. $w^5 \cdot w^0 \cdot w^{-7}$

3. $(3a^4 b^{-2})(a^5 b^{-3})$

4. $(4x^{-3}y^7)(-2x^2 y^2)$

5. $\dfrac{x^{-4}}{x^4}$

6. $\dfrac{15x^5 y^3}{3x^2 y^7}$

7. $\dfrac{x^5 \cdot x^{-4}}{x^{-3}}$

8. $x(3x^2 y^{-3})^2$

9. $(2w^{-2})^2 (5w^{-2})$

10. $x(5xy^{-2})^{-2}$

11. $\dfrac{m^{-9}s^{-8}}{m^{-4}s^3}$

12. $\dfrac{(3x^2y)^{-1}}{2xy^{-5}}$

13. $\dfrac{(3xy^{-2})^{-3}}{x}$

14. $\left[\dfrac{(3x^2y)^3}{3x^7y^9}\right]^2$

Hint: Since there are two sets of grouping symbols, work, as always, from the inside out to remove them.

15. $\left[\dfrac{(xy)^2}{x^{-1}}\right]^3$

16. $\left[\dfrac{(ab)^{-1}}{(a^{-2}b^3)^3}\right]^{-1}$ If you can do this one, you've got it made!

Solutions to Problems

1. $\dfrac{x^9y^{-6}}{x^{-3}y^2} = \dfrac{x^9x^3}{y^2y^6} = \dfrac{x^{9+3}}{y^{2+6}} = \dfrac{x^{12}}{y^8}$

2. $\left(\dfrac{2x^{-1}y^2}{x^{-3}}\right)^2 = \dfrac{2^2x^{-2}y^4}{x^{-6}}$

$= \dfrac{4x^6y^4}{x^2}$

$= 4x^{6-2}y^4$

$= 4x^4y^4$

UNIT 10

Review of Fractions: Addition and Subtraction

In the preceding units we discussed positive integral exponents as well as zero and negative exponents. Before learning about **fractional** exponents, we need to review some facts about fractions themselves.

In this unit we will review how fractions are added and subtracted. When you have completed the unit, you will be able to solve problems involving the addition and subtraction of fractions, without changing the terms to decimals.

Before we begin, we must be certain of several definitions:

Definition: Let a and b be integers with $b \neq 0$. Then $\dfrac{a}{b}$ is called a **rational number**

(generally referred to as a **fraction**). In the fraction $\dfrac{a}{b}$, a is the **numerator** and b is the **denominator**.

Definition: A **factor** is a number or letter that is being **multiplied**.

EXAMPLE 1

Consider: $3ax$. Since $3ax$ means $3 \cdot a \cdot x$, there are three factors: 3, a, and x.

> Definition: A fraction $\dfrac{a}{b}$ is said to be in **lowest terms** when all possible common factors **in the numerator and denominator** have been divided out.

EXAMPLE 2

Consider: $\dfrac{6}{15} = \dfrac{2 \cdot \cancel{3}}{5 \cdot \cancel{3}} = \dfrac{2}{5}$; $\dfrac{2}{5}$ is in lowest terms.

Recall my earlier comments on the **sign** of a fraction.

$$\frac{-a}{b} = \frac{a}{-b} = -\frac{a}{b}$$

These three are equivalent fractions! It is customary to place the negative sign either in the numerator or to the left of the fraction. Our preference is to place the negative sign in the numerator.

Let us add a note of caution before proceeding. Please do *not* change the fractions to decimals, or you will defeat the intention of this and the following unit, which is to enable you to work effectively with algebraic as well as numerical fractions.

ADDITION

> Rule: **Addition of Rationals with a Common Denominator**
>
> $$\frac{a}{b} + \frac{c}{b} = \frac{a+c}{b}$$

If the fractions have the same denominator, **add the numerators**. The denominator will remain the same.

EXAMPLE 3

Add: $\dfrac{1}{7} + \dfrac{2}{7}$.

Solution: $\dfrac{1}{7} + \dfrac{2}{7} = \dfrac{3}{7}$

EXAMPLE 4

Add: $\dfrac{2}{5}+\dfrac{4}{5}$.

Solution: $\dfrac{2}{5}+\dfrac{4}{5}=\dfrac{2+4}{5}=\dfrac{6}{5}$

>   Note: Leave the answer as an improper fraction; it is neither necessary nor advisable to change it to $1\dfrac{1}{5}$.

Unfortunately, few problems ever occur in which the fractions have the same denominator. The following rule allows us to add two fractions by finding a common denominator.

> Rule: **Addition of Rationals with Unlike Denominators**
>
> $$\frac{a}{b}+\frac{c}{d}=\frac{ad}{bd}+\frac{bc}{bd}=\frac{ad+bc}{bd}$$

Because of this rule, it is *not* necessary to find the least common denominator. However, the result may need to be reduced to lowest terms.

EXAMPLE 5

Add: $\dfrac{7}{8}+\dfrac{2}{3}$.

Solution: $\dfrac{7}{8}\times\dfrac{2}{3}=\dfrac{7\cdot3+8\cdot2}{8\cdot3}=\dfrac{21+16}{24}=\dfrac{37}{24}$

EXAMPLE 6

Add: $\dfrac{2}{3}+\dfrac{1}{5}$.

Solution: $\dfrac{2}{3}\times\dfrac{1}{5}=\dfrac{10+3}{15}=\dfrac{13}{15}$

Now try two problems yourself.

Problem 1

Add: $\dfrac{4}{7} + \dfrac{5}{6}$.

Solution:

Problem 2

Add: $\dfrac{-2}{3} + \dfrac{2}{5}$.

Solution:

The advantage of the rule given above is that for simple fractions the addition can be performed in one step. However, it can be used only when adding two fractions; if there are more, the fractions must be added two at a time with this definition.

SUBTRACTION

Rule: **Subtraction of Rationals with Unlike Denominators**

$$\frac{a}{b} - \frac{c}{d} = \frac{ad - bc}{bd}$$

Note that the only difference between this definition and the definition for the addition of rationals is the minus sign.

But be careful—the first term of the numerator must be the product found by multiplying diagonally to the right and **down** followed by the **minus sign** since it is a subtraction problem.

$$\frac{a}{b} \searrow \frac{c}{d} = \frac{ad -}{\underline{\hspace{2cm}}}$$

EXAMPLE 7

Subtract: $\dfrac{3}{5} - \dfrac{2}{3}$.

Solution: $\dfrac{3}{5} \searrow \dfrac{2}{3} = \dfrac{9-10}{15} = \dfrac{-1}{15}$

EXAMPLE 8

Subtract: $\dfrac{-5}{11} - \dfrac{3}{4}$.

Solution: $\dfrac{-5}{11} \searrow \dfrac{3}{4} = \dfrac{-20-33}{44} = \dfrac{-53}{44}$

EXAMPLE 9

Subtract: $\dfrac{1}{3} - \dfrac{1}{2}$.

Solution: $\dfrac{1}{3} \searrow \dfrac{1}{2} = \dfrac{2-3}{6} = \dfrac{-1}{6}$

EXAMPLE 10

Solve: $\dfrac{1}{6} - \dfrac{-2}{3}$.

Solution: $\dfrac{1}{6} \searrow \dfrac{-2}{3} = \dfrac{3-(-12)}{18} = \dfrac{3+12}{18} = \dfrac{15}{18} = \dfrac{\cancel{3} \cdot 5}{\cancel{3} \cdot 6} = \dfrac{5}{6}$

Note: After using the rule, our solution was not in the lowest terms and had to be reduced in the final step.

Try a couple of subtraction problems on your own. If you have forgotten the rules for adding and subtracting signed numbers, quickly review Unit 2 before proceeding.

Problem 3

Solve: $\dfrac{5}{3} - \dfrac{1}{7}$.

Solution:

Problem 4

Solve: $\dfrac{-3}{4} - \dfrac{2}{5}$.

Solution:

With algebraic fractions the entire process becomes even easier.

EXAMPLE 11

Add: $\dfrac{1}{x} + \dfrac{1}{y}$.

Solution: $\dfrac{1}{x} + \dfrac{1}{y} = \dfrac{y+x}{xy}$

EXAMPLE 12

Add: $\dfrac{2x}{y} - \dfrac{z}{3}$.

Solution: $\dfrac{2x}{y} - \dfrac{z}{3} = \dfrac{6x - yz}{3y}$

The next example illustrates how to add or subtract when only one of the terms is a fraction.

EXAMPLE 13

Subtract: $x - \dfrac{1}{2}$.

Solution: $x - \dfrac{1}{2} = \dfrac{x}{1} - \dfrac{1}{2} = \dfrac{x}{1} \searrow \dfrac{1}{2} = \dfrac{2x-1}{2}$

First the x was rewritten as $\dfrac{x}{1}$, and then the two fractions were subtracted.

We will conclude this unit with one final example.

EXAMPLE 14

Subtract: $\dfrac{x+2}{5} - \dfrac{x-1}{7}$.

Solution: $\dfrac{x+2}{5} \searrow \dfrac{x-1}{7} = \dfrac{7(x+2) - 5(x-1)}{35}$

$\qquad = \dfrac{7x + 14 - 5x + 5}{35}$ Note: The sign changed to +5.

$\qquad = \dfrac{2x + 19}{35}$

You should now be able to add and subtract rationals (fractions). Before beginning the next unit, you should solve the following problems involving rationals. Reduce the answers to lowest terms, but *do not change them to decimals.*

EXERCISES

Express as a single fraction in lowest terms.

1. $\dfrac{2}{11} + \dfrac{1}{11}$

2. $\dfrac{7}{10} - \dfrac{9}{10}$

3. $\dfrac{7}{9} + \dfrac{1}{5}$

4. $\dfrac{1}{x} + 5$ Hint: Rewrite 5 as $\dfrac{5}{1}$.

5. $3 - \dfrac{5}{w}$

6. $7 + \dfrac{2}{x}$

7. $\dfrac{2}{9} - \dfrac{-1}{10}$

8. $\dfrac{11}{t} + \dfrac{7}{r}$

9. $\dfrac{1}{x} + \dfrac{1}{x}$

10. $\dfrac{10}{x+1} + \dfrac{3}{x+1}$

11. $\dfrac{5}{a} - \dfrac{4}{a}$

12. $\dfrac{-s}{9} + \dfrac{k}{10}$

13. $\dfrac{x}{2} - \dfrac{x}{5}$

14. $\dfrac{x+1}{2} - \dfrac{3}{5}$

15. $\dfrac{x-1}{3} + \dfrac{x+1}{2}$

16. $\dfrac{x+2}{2} - \dfrac{x+3}{3}$

Solutions to Problems

1. $\dfrac{4}{7} + \dfrac{5}{6} = \dfrac{4 \cdot 6 + 7 \cdot 5}{7 \cdot 6} = \dfrac{24 + 35}{42} = \dfrac{59}{42}$

2. $\dfrac{-2}{3} + \dfrac{2}{5} = \dfrac{-2 \cdot 5 + 3 \cdot 2}{3 \cdot 5} = \dfrac{-10 + 6}{15} = \dfrac{-4}{15}$

3. $\dfrac{5}{3} - \dfrac{1}{7} = \dfrac{5 \cdot 7 - 3 \cdot 1}{3 \cdot 7} = \dfrac{35 - 3}{21} = \dfrac{32}{21}$

4. $\dfrac{-3}{4} - \dfrac{2}{5} = \dfrac{-3 \cdot 5 - 4 \cdot 2}{4 \cdot 5} = \dfrac{-15 - 8}{20} = \dfrac{-23}{20}$

UNIT 11

Review of Fractions: Multiplication and Division; Complex Fractions

Here, as in Unit 10, we will review some facts about **fractions** before we study **fractional exponents**. When you have completed this unit, you will be able to solve problems involving the multiplication and division of fractions, without changing the terms to decimals.

MULTIPLICATION

> Rule: **Multiplication of Rationals**
>
> $$\frac{a}{b} \cdot \frac{c}{d} = \frac{ac}{bd}$$

Multiply numerators together, and multiply denominators together.

Note: 1. You do not need a common denominator.
2. You should factor and cancel as soon as possible.

EXAMPLE 1

Multiply: $\dfrac{2}{3} \cdot \dfrac{4}{5}$.

Solution: $\dfrac{2}{3} \cdot \dfrac{4}{5} = \dfrac{2 \cdot 4}{3 \cdot 5} = \dfrac{8}{15}$

EXAMPLE 2

Multiply: $\dfrac{4}{7} \cdot \dfrac{35}{12}$.

Solution: $\dfrac{\overset{1}{\cancel{4}}}{\underset{1}{\cancel{7}}} \cdot \dfrac{\overset{5}{\cancel{35}}}{\underset{3}{\cancel{12}}} = \dfrac{5}{3}$

EXAMPLE 3

Multiply: $\dfrac{12}{15} \cdot \dfrac{5}{18}$.

Solution: $\dfrac{\overset{2}{\cancel{12}}}{\underset{3}{\cancel{15}}} \cdot \dfrac{\overset{1}{\cancel{5}}}{\underset{3}{\cancel{18}}} = \dfrac{2 \cdot 1}{3 \cdot 3} = \dfrac{2}{9}$

DIVISION

> Rule: **Division of Rationals**
>
> $$\frac{a}{b} \div \frac{e}{f} = \frac{a}{b} \cdot \frac{f}{e} = \frac{af}{be}$$

When dividing by a fraction, we invert the second fraction (the divisor) and multiply. Or, another way of saying it, we multiply by the reciprocal of the divisor.

EXAMPLE 4

Divide: $\dfrac{2}{3} \div \dfrac{-3}{8}$.

Solution: $\dfrac{2}{3} \div \dfrac{-3}{8} = \dfrac{2}{3} \cdot \dfrac{8}{-3} = \dfrac{16}{-9} = \dfrac{-16}{9}$

As stated in the previous unit, with algebraic fractions the entire process becomes even easier.

EXAMPLE 5

Multiply: $\dfrac{a}{b} \cdot \dfrac{5}{a}$.

Solution: $\dfrac{a}{b} \cdot \dfrac{5}{a} = \dfrac{\cancel{a} \cdot 5}{b \cdot \cancel{a}} = \dfrac{5}{b}$

EXAMPLE 6

Divide: $\dfrac{x+1}{3} \div 2$.

Solution: $\dfrac{x+1}{3} \div 2 = \dfrac{x+1}{3} \cdot \dfrac{1}{2} = \dfrac{(x+1) \cdot 1}{3 \cdot 2} = \dfrac{x+1}{6}$

Here are four problems for you.

Problem 1

Simplify: $\dfrac{2}{-5} \cdot \dfrac{30}{8}$.

Solution:

Problem 2

Simplify: $\dfrac{2}{-5} \div \dfrac{30}{8}$.

Solution:

Problem 3

Simplify: $\dfrac{-12}{20} \div \dfrac{10}{-16}$.

Solution:

Problem 4

Divide: $\dfrac{w+x}{2} \div \dfrac{1}{2}$.

Solution:

COMPLEX FRACTIONS

We must now consider **complex fractions**—fractions in which there are one or more fractions in the numerator or denominator, or in both. Remember that in a complex fraction, as in a simple fraction, the horizontal bar means simply that we should divide. For example:

$$\frac{12}{2} \text{ means 12 divided by 2} \quad (12 \div 2)$$

Therefore, when we are faced with a complex fraction, the first thing we do is to rewrite it as a simple division problem. Remember that if the numerator or the denominator, or both, have more than one term, you will need to enclose them in parentheses when rewriting the complex fraction as a division problem.

EXAMPLE 7

Simplify this complex fraction: $\dfrac{\frac{2}{5}}{w}$.

Solution: $\dfrac{\frac{2}{5}}{w} = \dfrac{2}{5} \div w$ $\quad$ Rewrite, using a division sign.

$$= \frac{2}{5} \cdot \frac{1}{w}$$

$$= \frac{2}{5w}$$

EXAMPLE 8

Simplify: $\dfrac{\frac{1}{2} + \frac{2}{3}}{\frac{5}{4}}$.

Solution: $\dfrac{\frac{1}{2}+\frac{2}{3}}{\frac{5}{4}} = \left(\dfrac{1}{2}+\dfrac{2}{3}\right) \div \dfrac{5}{4}$ Rewrite, using a division sign.

$= \left(\dfrac{3+4}{6}\right) \div \dfrac{5}{4}$ Add fractions in parentheses.

$= \dfrac{7}{6} \cdot \dfrac{4}{5}$ Multiply by the reciprocal of the divisor.

$= \dfrac{7}{\cancel{6}_3} \cdot \dfrac{\cancel{4}^2}{5}$

$= \dfrac{14}{15}$

EXAMPLE 9

Simplify: $\dfrac{3+\frac{1}{x}}{2}$.

Solution: $\dfrac{3+\frac{1}{x}}{2} = \left(3+\dfrac{1}{x}\right) \div 2$ Rewrite, using a division sign.

$= \left(\dfrac{3}{1}+\dfrac{1}{x}\right) \div \dfrac{2}{1}$

$= \dfrac{3x+1}{x} \cdot \dfrac{1}{2}$ Add fractions in parentheses.

$= \dfrac{3x+1}{2x}$

Can you do this one?

Problem 5

Simplify: $\dfrac{\frac{3}{x}}{1+\frac{7}{x}}$

Solution:

Here is one more example to end the unit.

EXAMPLE 10

Simplify: $\dfrac{5-\dfrac{3}{x}}{x}$.

Solution:

$$\frac{5-\dfrac{3}{x}}{x}=\left(\frac{5}{1}-\frac{3}{x}\right)\div x$$

$$=\left(\frac{5x-3}{x}\right)\cdot\frac{1}{x}$$

$$=\frac{(5x-3)\cdot 1}{x\cdot x}$$

$$=\frac{5x-3}{x^2}$$

You should now be able to multiply and divide fractions. You should also be able to simplify complex fractions in which there are fractions in the numerator and/or denominator.

Before beginning the next unit, do the following exercises, reducing the answers to the lowest terms *without converting to decimals*.

EXERCISES

Simplify:

1. $\dfrac{3}{8}\cdot\dfrac{32}{15}$

2. $\dfrac{5}{18}\div\dfrac{3}{14}$

3. $\dfrac{3}{4}\cdot\dfrac{-8}{9}$

4. $\dfrac{9}{14}\div\dfrac{5}{21}$

5. $\dfrac{-8}{9}\div\dfrac{12}{-7}$

6. $\left(\dfrac{2}{x}\cdot\dfrac{x}{5}\right)\div w$

7. $\dfrac{\dfrac{3}{10}}{\dfrac{1}{10}}$

8. $\dfrac{3-\dfrac{2}{5}}{3+\dfrac{2}{5}}$

9. $\dfrac{1-\dfrac{1}{3}}{\dfrac{5}{6}}$

10. $\dfrac{\dfrac{a}{2}-\dfrac{3}{5}}{2}$

11. $7\div\left(\dfrac{x-1}{4}\right)$

12. $\dfrac{\dfrac{2}{x}-5}{x}$

13. $\dfrac{9-\dfrac{x}{4}}{\dfrac{1}{2}}$

15. $\dfrac{\dfrac{a}{b}+2}{\dfrac{a}{b}+1}$

14. $\dfrac{x+\dfrac{x+2}{2}}{\dfrac{x}{2}}$

Solutions to Problems

1. $\dfrac{\cancel{2}}{\cancel{-5}}\cdot\dfrac{\cancel{30}^{-6}}{\cancel{8}_4}=\dfrac{\cancel{-6}^{-3}}{\cancel{4}_2}=\dfrac{-3}{2}$

 or

 $\dfrac{2}{-5}\cdot\dfrac{30}{8}=\dfrac{2\cdot30}{-5\cdot8}=\dfrac{\cancel{60}^3}{\cancel{-40}_{-2}}=\dfrac{3}{-2}=\dfrac{-3}{2}$

2. $\dfrac{2}{-5}\div\dfrac{30}{8}=\dfrac{\cancel{2}}{-5}\cdot\dfrac{8}{\cancel{30}_{15}}=\dfrac{8}{-5\cdot15}=\dfrac{8}{-75}=\dfrac{-8}{75}$

3. $\dfrac{-12}{20}\div\dfrac{10}{-16}=\dfrac{\cancel{-12}^{-6}}{\cancel{20}_5}\cdot\dfrac{\cancel{-16}^{-4}}{\cancel{10}_5}=\dfrac{-6\cdot-4}{5\cdot5}=\dfrac{24}{25}$

4. $\dfrac{w+x}{2}\div\dfrac{1}{2}=\dfrac{w+x}{\cancel{2}}\cdot\dfrac{\cancel{2}}{1}=w+x$

5. $\dfrac{\dfrac{3}{x}}{1+\dfrac{7}{x}}=\dfrac{3}{x}\div\left(1+\dfrac{7}{x}\right)$

 $=\dfrac{3}{x}\div\left(\dfrac{1}{1}+\dfrac{7}{x}\right)$

 $=\dfrac{3}{x}\div\left(\dfrac{x+7}{x}\right)$

 $=\dfrac{3}{x}\cdot\left(\dfrac{x}{x+7}\right)$

 $=\dfrac{3}{\cancel{x}}\cdot\left(\dfrac{\cancel{x}}{x+7}\right)$

 $=\dfrac{3}{x+7}$

UNIT 12

Square Roots and Radicals

In this unit you will learn what is meant by the root of a number or algebraic expression and the vocabulary used with radicals. When you have completed the unit, you will be able to multiply, divide, add, and subtract radical expressions, both numerical and algebraic.

ROOTS

The **square root** of a number or expression is one of its *two equal factors*. For example, +5 is a square root of 25 since $(+5)(+5) = 25$. Also, −5 is a square root of 25 since $(-5)(-5) = 25$.

A positive number has two square roots, which are opposites of each other. To indicate both square roots, the symbol $\pm$ may be used. Thus the square roots of 49 are ± 7.

The **cube root** of a number or expression is one of its *three equal factors*. For example, 2 is the cube root of 8 since $(2)(2)(2) = 8$, and −3 is the cube root of −27 since $(-3)(-3)(-3) = -27$, and $2x$ is the cube root of $8x^3$ since $(2x)(2x)(2x) = 8x^3$.

The fourth root of a number is one of its *four equal factors*, and so on for fifth roots, etc.

EXAMPLE 1   The square roots of 36 are ± 6 since $(6)(6) = 36$ and $(-6)(-6) = 36$.

EXAMPLE 2   The cube root of −8 is −2 since $(-2)(-2)(-2) = -8$.

EXAMPLE 3   The fifth root of 1 is 1 since $(1)(1)(1)(1)(1) = 1$.

EXAMPLE 4   The square root of −16 does not exist in the real number system because there is no real number such that when it is squared will be a negative number.

- -

From the four examples we should note the following:

1. A positive number has two square roots, which are opposites of each other.

2. A positive number has only one cube root.

3. The cube root of a negative number exists and is negative.

4. The square root of a negative number does not exist in the set of real numbers.

Try the next four problems on your own.

Problem 1   Find the square roots of 64.

Problem 2   Find the square roots of 100.

Problem 3   Find the cube root of –1.

Problem 4   Find the square root of 0.

- -

RADICALS

A **radical** is an indicated root of a number or expression such as $\sqrt{5}$, $\sqrt[3]{8x}$, and $\sqrt[4]{7x^3}$. The symbol, $\sqrt{\ }$, is called a **radical sign**.

The **radicand** is the number or expression under the radical sign. In the above examples the radicands are 5, 8x, and $7x^3$.

The **index** is the small number written above and to the left of the radical sign. The index indicates which root is to be taken. In square roots, the index 2 is not written. In the above examples the indices are 2, 3, and 4.

The **principal square root** of a number is its positive square root. The radical sign is used to indicate the principal square root. Thus the square roots of 49 are ±7, but $\sqrt{49}$ = principal square root of 49 = +7. To indicate the negative square root of a number, a negative sign is placed before the symbol. $-\sqrt{16} = -4$.

EXAMPLE 5

Find $\sqrt{4}$.

- -

Solution: The problem is asking for the principal square root of 4, which is the positive root only.

Answer: $\sqrt{4} = 2$

- -

EXAMPLE 6

Find $\sqrt[3]{-125}$.

Solution: The problem is asking for the cube root of −125.

Answer: $\sqrt[3]{-125} = -5$ since $(-5)(-5)(-5) = -125$

Problem 5   Find $\sqrt[3]{64}$.

Problem 6   Find $\sqrt[5]{32}$.

Problem 7   Find $\sqrt[4]{81}$.

Numbers such as $\sqrt{5}$, $\sqrt{23}$, and $\sqrt[3]{17}$, where the radicand is not a perfect square or cube, are irrational numbers. If a decimal approximation is needed, it can be found by using a calculator; otherwise the radical is left as written. The intent in this and later units is to learn to work with radical expressions, not their decimal approximations.

SQUARE ROOTS OF SQUARES

For the remainder of this book, we will assume that all letters—a, b, c, . . . , x, y, z—that appear under a radical sign represent positive real numbers unless otherwise specified. Thus $\sqrt{x^2} = x$ since x is assumed to be positive and is the principal square root. In later mathematics courses, you will need to consider the possibility that x is negative, and the principal square root becomes $\sqrt{x^2} = |x|$.

In the following examples the algebraic expression under the radical sign, the radicand, is a square. To find the square root of the variable part of a monomial, which is a square, we need only keep the base the same and take one half of the exponent.

EXAMPLE 7 $\sqrt{y^2} = y$ since $y \cdot y = y^2$

EXAMPLE 8 $\sqrt{9w^2} = 3w$ since $(3w)(3w) = (3w)^2 = 3^2 w^2 = 9w^2$

EXAMPLE 9 $\sqrt{a^2 c^2} = ac$ since $(ac)(ac) = (ac)^2 = a^2 c^2$

EXAMPLE 10 $\sqrt{x^6} = x^3$ since $x^3 \cdot x^3 = x^6$

EXAMPLE 11 $\sqrt{z^{10}} = z^5$ since $z^5 \cdot z^5 = z^{10}$

OTHER ROOTS

This same idea can be modified for cube roots of perfect cubes and higher roots of higher powers. To find the cube root of the variable part of a monomial that is a perfect cube, we keep the same base and take one third of the exponent. For a fourth root of a fourth power, we take one fourth of the exponent, and so on.

EXAMPLE 12 $\sqrt[3]{x^3} = x$ since $x \cdot x \cdot x = x^3$

EXAMPLE 13 $\sqrt[3]{8y^{12}} = 2y^4$ since $(2y^4)(2y^4)(2y^4) = (2y^4)^3 = 8y^{12}$

EXAMPLE 14 $\sqrt[4]{x^4} = 4$ since $x \cdot x \cdot x \cdot x = x^4$

EXAMPLE 15 $\sqrt[4]{z^{12}} = z^3$ since $(z^3)(z^3)(z^3)(z^3) = (z^3)^4 = z^{12}$

SIMPLIFYING SQUARE ROOTS OF POWERS

This section deals with how to simplify an expression like $\sqrt{x^3}$, where the expression under the radical is not itself a square but does have square factors. We can use the rule for multiplying square roots in reverse. First rewrite the expression as a product of the greatest square factor times the remaining factors. To simplify, take the square root of the square.

EXAMPLE 16 $\sqrt{x^3} = \sqrt{x^2 \cdot x} = \sqrt{x^2} \cdot \sqrt{x} = x\sqrt{x}$

EXAMPLE 17 $\sqrt{12x} = \sqrt{4 \cdot 3x} = \sqrt{4} \cdot \sqrt{3x} = 2\sqrt{3x}$

EXAMPLE 18 $\sqrt{50w^5} = \sqrt{25 \cdot 2 \cdot w^4 \cdot w} = \sqrt{25w^4} \cdot \sqrt{2w} = 5w^2\sqrt{2w}$

Problem 8 $\sqrt{z^5}$

Problem 9 $\sqrt{48y^7}$

MULTIPLICATION AND DIVISION

Now that you are able to simplify square radicands, we will consider multiplying and dividing square roots.

$$\boxed{\text{Rule:} \quad \sqrt{a} \cdot \sqrt{b} = \sqrt{ab}}$$

In words, the rule states that the product of square roots is the square root of the product of the radicands. If the resulting new radicand is a square, it should be simplified as illustrated in the following examples.

EXAMPLE 19 $\sqrt{2} \cdot \sqrt{5} = \sqrt{2 \cdot 5} = \sqrt{10}$

EXAMPLE 20 $\sqrt{5} \cdot \sqrt{x} = \sqrt{5 \cdot x} = \sqrt{5x}$

EXAMPLE 21 $\sqrt{a^3} \cdot \sqrt{a} = \sqrt{a^3 \cdot a} = \sqrt{a^4} = a^2$

EXAMPLE 22 $\sqrt{2x} \cdot \sqrt{8x} = \sqrt{2 \cdot 8 \cdot x \cdot x} = \sqrt{16x^2} = 4x$

EXAMPLE 23 $2\sqrt{3} \cdot 5\sqrt{15} = 2 \cdot 5\sqrt{3 \cdot 15} = 10\sqrt{45} = 10\sqrt{9 \cdot 5} = 10 \cdot 3\sqrt{5} = 30\sqrt{5}$

EXAMPLE 24 $\sqrt{5}\left(\sqrt{3} + \sqrt{w}\right) = \sqrt{5 \cdot 3} + \sqrt{5 \cdot w} = \sqrt{15} + \sqrt{5w}$

A similar rule applies for dividing square root radicals.

$$\text{Rule:} \quad \frac{\sqrt{a}}{\sqrt{b}} = \sqrt{\frac{a}{b}}$$

In words, the rule states that the quotient of two square roots is equal to the square root of the quotient of the radicands. One example should be sufficient.

EXAMPLE 25 $\dfrac{\sqrt{27x^3}}{\sqrt{3x}} = \sqrt{\dfrac{27x^3}{3x}} = \sqrt{9x^2} = 3x$

It's time for you to try a few problems on your own. Perform the indicated operations and simplify the square roots of any squares.

Problem 10 $\sqrt{49g^2}$

Problem 11 $\sqrt{20w} \cdot \sqrt{5w^3}$

Problem 12 $\sqrt{x} \cdot \sqrt{x}$

Problem 13 $3\sqrt{5} \cdot 4\sqrt{12}$

Problem 14 $\dfrac{\sqrt{45x^7}}{\sqrt{5x^3}}$

RATIONALIZING MONOMIAL DENOMINATORS

It often happens when dividing radicals that the answer contains a radical in the denominator. For example, $\dfrac{\sqrt{28}}{\sqrt{8}} = \sqrt{\dfrac{28}{8}} = \sqrt{\dfrac{7}{2}} = \dfrac{\sqrt{7}}{\sqrt{2}}$. Neither a radical in the denominator nor, equivalently, a fraction or a decimal inside the radicand are considered to be in simplest form. To simplify such an answer, we **rationalize the denominator** as follows.

> **Rule: Rationalizing a Monomial Denominator**
>
> $$\frac{\sqrt{a}}{\sqrt{b}} = \frac{\sqrt{a}}{\sqrt{b}} \cdot \frac{\sqrt{b}}{\sqrt{b}} = \frac{\sqrt{ab}}{b}$$

In words, the rule says to multiply both the numerator and denominator of the expression by the radical in the denominator. The result will have a number in the denominator.

EXAMPLE 26 $\quad \dfrac{\sqrt{7}}{\sqrt{2}} = \dfrac{\sqrt{7}}{\sqrt{2}} \cdot \dfrac{\sqrt{2}}{\sqrt{2}} = \dfrac{\sqrt{14}}{2}$

Try it yourself. Express the following in simplest form with rational denominators.

Problem 15 $\quad \dfrac{\sqrt{6x}}{\sqrt{5}}$

Problem 16 $\quad \dfrac{\sqrt{15x}}{\sqrt{6}}$

ADDITION AND SUBTRACTION

To add or subtract radicals, the radicands must be the same. Then combine like terms by adding or subtracting the coefficients of the common radicals.

EXAMPLE 27 $\quad 2\sqrt{7} + 3\sqrt{7} = (2+3)\sqrt{7} = 5\sqrt{7}$

EXAMPLE 28 $\quad 9\sqrt{2x} - \sqrt{2x} = 9\sqrt{2x} - 1\sqrt{2x} = (9-1)\sqrt{2x} = 8\sqrt{2x}$

The final example requires that the radicals be simplified before attempting to do the addition.

EXAMPLE 29

Simplify: $\quad 2\sqrt{32w} + \sqrt{18w}$.

Solution: $\quad 2\sqrt{32w} + \sqrt{18w} = 2\sqrt{16 \cdot 2w} + \sqrt{9 \cdot 2w}$

$$= 2 \cdot 4\sqrt{2w} + 3\sqrt{2w}$$

$$= 8\sqrt{2w} + 3\sqrt{2w}$$

$$= (8+3)\sqrt{2w}$$

$$= 11\sqrt{2w}$$

You should now be able to simplify, add, subtract, multiply, and divide square root radicals, both numerical and algebraic expressions.

Before continuing on to the next unit, you should simplify the following radicals. Do not use decimal approximations but instead work with simplifying the expressions written as radicals.

Problem 17 $3\sqrt{5} + 12\sqrt{5}$

Problem 18 $10\sqrt{3y} - 7\sqrt{3y}$

Problem 19 $4\sqrt{18x} - 5\sqrt{2x}$

EXERCISES

Simplify:

1. $\sqrt{5} \cdot \sqrt{20}$

2. $\sqrt{75}$

3. $\sqrt{2a} \cdot \sqrt{3b}$

4. $\sqrt{0}$

5. $\sqrt{3} \cdot \sqrt{6}$

6. $\sqrt{64t^2}$

7. $\sqrt{w^4}$

8. $\sqrt{45}$

9. $\sqrt{\dfrac{25}{x^2}}$

10. $\dfrac{\sqrt{20}}{\sqrt{5}}$

11. $11\sqrt{2} + 3\sqrt{2}$

12. $\sqrt{3x} \cdot \sqrt{3x}$

13. $7\sqrt{40} - 2\sqrt{10}$

14. $\sqrt{12y^8}$

15. $\sqrt{50x^4}$

16. $\sqrt{3}(\sqrt{2} + 1)$

17. $\sqrt{5}(\sqrt{5} + \sqrt{3})$

18. $\sqrt{2a^2c} \cdot \sqrt{2ac}$

19. $\dfrac{\sqrt{48}}{\sqrt{18}}$

20. $\dfrac{\sqrt{21x}}{\sqrt{15x^3}}$

Solutions to Problems

1. 8 and −8, often written as ±8

2. ±10

3. −1

4. 0

5. $\sqrt[3]{64} = 4$

6. $\sqrt[5]{32} = 2$

7. $\sqrt[4]{81} = 3$

8. $\sqrt{z^5} = \sqrt{z^4 \cdot z} = \sqrt{z^4} \cdot \sqrt{z} = z^2\sqrt{z}$

9. $\sqrt{48y^7} = \sqrt{16 \cdot 3 \cdot y^6 \cdot y} = \sqrt{16y^6} \cdot \sqrt{3y} = 4y^3\sqrt{3y}$

10. $\sqrt{49g^2} = 7g$

11. $\sqrt{20w} \cdot \sqrt{5w^3} = \sqrt{20w \cdot 5w^3} = \sqrt{100w^4} = 10w^2$

12. $\sqrt{x} \cdot \sqrt{x} = \sqrt{x^2} = x$

13. $3\sqrt{5} \cdot 4\sqrt{12} = 3 \cdot 4\sqrt{5 \cdot 12} = 12\sqrt{60} = 12\sqrt{4 \cdot 15} = 12 \cdot 2\sqrt{15} = 24\sqrt{15}$

14. $\dfrac{\sqrt{45x^7}}{\sqrt{5x^3}} = \sqrt{\dfrac{45x^7}{5x^3}} = \sqrt{9x^4} = 3x^2$

15. $\dfrac{\sqrt{6x}}{\sqrt{5}} = \dfrac{\sqrt{6x}}{\sqrt{5}} \cdot \dfrac{\sqrt{5}}{\sqrt{5}} = \dfrac{\sqrt{30x}}{5}$

16. $\dfrac{\sqrt{15x}}{\sqrt{6}} = \dfrac{\sqrt{15x}}{\sqrt{6}} \cdot \dfrac{\sqrt{6}}{\sqrt{6}} = \dfrac{\sqrt{90x}}{6} = \dfrac{\sqrt{9 \cdot 10x}}{6} = \dfrac{3\sqrt{10x}}{6} = \dfrac{\cancel{3}\sqrt{10x}}{\cancel{6}_2} = \dfrac{\sqrt{10x}}{2}$

17. $3\sqrt{5} + 12\sqrt{5} = (3 + 12)\sqrt{5} = 15\sqrt{5}$

18. $10\sqrt{3y} - 7\sqrt{3y} = (10 - 7)\sqrt{3y} = 3\sqrt{3y}$

19. $4\sqrt{18x} - 5\sqrt{2x} = 4\sqrt{9 \cdot 2x} - 5\sqrt{2x}$

    $\qquad\qquad\qquad = 4 \cdot 3\sqrt{2x} - 5\sqrt{2x}$

    $\qquad\qquad\qquad = 12\sqrt{2x} - 5\sqrt{2x}$

    $\qquad\qquad\qquad = (12 - 5)\sqrt{2x}$

    $\qquad\qquad\qquad = 7\sqrt{2x}$

UNIT 13

Fractional Exponents

In Units 10 and 11 we reviewed how to add, subtract, multiply, and divide fractions. Then in Unit 12 you learned what is meant by the root of a number or algebraic expression and the vocabulary used with radicals. Now we are ready to proceed to handling the most difficult exponents, fractional exponents.

First recall how we define integral exponents:

$$
\begin{array}{ll}
\text{Positive Integer:} & b^n = \underbrace{b \cdot b \cdot b \cdot \ldots \cdot b}_{n \text{ factors}} \\[2em]
\text{Zero:} & b^0 = 1 \\[1em]
\text{Negative:} & b^{-n} = \dfrac{1}{b^n} \quad \text{and} \quad \dfrac{1}{b^{-n}} = b^n
\end{array}
$$

Now we will define a fractional exponent.

$$
\boxed{
\begin{array}{l}
\text{Definition:} \quad \textbf{Fractional Exponent} \\[1em]
b^{n/d} = \left(\sqrt[d]{b}\right)^n = \sqrt[d]{b^n} \\[1em]
\text{If } d \text{ is even, } b \text{ must be nonnegative.}
\end{array}
}
$$

Using the vocabulary introduced in Unit 12, the letter d is called the **index** of the radical and b is called the **radicand**.

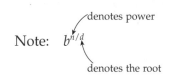

Note: $b^{n/d}$ — denotes power, denotes the root

Note that two forms of the definition are given.

$$b^{n/d} = \left(\sqrt[d]{b}\right)^n = \sqrt[d]{b^n}$$

numerical algebraic

The first form is useful in numerical calculations, provided that the dth root of b is a known integer. It is then convenient to take the root before raising to the nth power in order to work with smaller numbers.

The second form is the more common way of rewriting algebraic expressions with fractional exponents.

As the two forms suggest, we can do either the root or the power first. Here we will consider some numerical bases first.

Fractional Exponent with Numerical Problems

power   root

$$b^{n/d} = \left(\sqrt[d]{b}\right)^n$$

numerical

EXAMPLE 1 $8^{1/3} = \left(\sqrt[3]{8}\right)^1 = (2)^1 = 2$

EXAMPLE 2 $4^{1/2} = \left(\sqrt{4}\right)^1 = (2)^1 = 2$

Remember that the 2 is not usually written for square roots.

EXAMPLE 3 $4^{3/2} = \left(\sqrt{4}\right)^3 = (2)^3 = 8$

EXAMPLE 4 $27^{2/3} = \left(\sqrt[3]{27}\right)^2 = (3)^2 = 9$

EXAMPLE 5 $8^{4/3} = \left(\sqrt[3]{8}\right)^4 = (2)^4 = 16$

EXAMPLE 6 $9^{-1/2} = \left(\sqrt{9}\right)^{-1} = (3)^{-1} = \dfrac{1}{3}$

or, alternatively,

$$= \frac{1}{9^{1/2}} = \frac{1}{\sqrt{9}} = \frac{1}{3}$$

It is important that you learn the meaning of, and be able to work with, fractional exponents. Punching numbers into a calculator defeats this purpose. So for now try to simplify the following numerical problems with fractional exponents without using a calculator. You should be able to find the various roots easily.

Problem 1 $32^{1/5} =$

Problem 2 $(-1)^{1/3} =$

Problem 3 $4^{5/2} =$

Problem 4 $64^{2/3} =$

Problem 5 $81^{3/4} =$

Problem 6 $(-8)^{2/3} =$

And now for some problems where the root is not a rational number. When the root is not a rational number, as in the next two examples, the expression is rewritten as a radical. Although the two expressions are equivalent, the radical form is considered to be the simplified form. For example, $2^{1/2}$ and $\sqrt{2}$ are equivalent, but $\sqrt{2}$ is considered to be the simplified form.

EXAMPLE 7   Rewrite as a radical: $5^{2/3} = \sqrt[3]{5^2} = \sqrt[3]{25}$

EXAMPLE 8   Rewrite as a radical: $7^{1/5} = \sqrt[5]{7}$

Fractional Exponent with Algebraic Expressions

$$\overset{\text{power}\quad\text{root}}{b^{n/d} = \sqrt[d]{b^n}}$$

algebraic

EXAMPLE 9   Rewrite as a radical: $x^{2/3} = \sqrt[3]{x^2}$

EXAMPLE 10   Rewrite as a radical: $y^{2/5} = \sqrt[5]{y^2}$

The only time any simplifying can be done is if n/d is an improper fraction. Given an algebraic expression with an improper fractional exponent, rewrite the expression as a radical and simplify as explained in Unit 12. The following examples illustrate the procedure.

EXAMPLE 11 $\quad x^{3/2} = \sqrt{x^3} = \sqrt{x^2 \cdot x} = x\sqrt{x}$

The final answer is preferred to $\sqrt{x^3}$.

EXAMPLE 12 $\quad x^{7/5} = \sqrt[5]{x^7} = \sqrt[5]{x^5 \cdot x^2} = x\sqrt[5]{x^2}$

EXAMPLE 13 $\quad x^{4/3} = \sqrt[3]{x^4} = \sqrt[3]{x^3 \cdot x} = x\sqrt[3]{x}$

EXAMPLE 14 $\quad x^{5/2} = \sqrt{x^5} = \sqrt{x^4 \cdot x} = x^2\sqrt{x}$

ALGEBRA AND THE CALCULATOR (Optional)

In some instances you may need a decimal approximation for a radical. There is a calculator key for square roots, which makes these problems relatively easy. But what about examples such as $5^{2/3}$ and $-10^{-1/5}$? The $\boxed{\wedge}$ key is used to denote exponentiation. In the next example the keystrokes and screen are shown for finding the decimal approximation.

EXAMPLE 15 Using a calculator, find the decimal approximation for $5^{2/3}$.

Solution: To find the answer, press $\boxed{5}$ $\boxed{\wedge}$ $\boxed{\text{ALPHA}}$ $\boxed{\text{Y=}}$ $\boxed{1}$ $\boxed{2}$ $\boxed{)}$ $\boxed{3}$ $\boxed{)}$ $\boxed{\text{ENTER}}$.

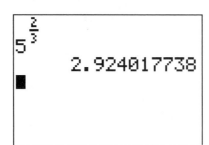

Answer: 2.924, correct to three decimal places

The calculator can do higher order roots than square roots, but the symbol is hidden in the Math menu. To find the symbol for the cube root or to specify the index in the root, press $\boxed{\text{MATH}}$, select 4 for a cube root and 5 for a 4th root or higher. To use $\sqrt[x]{}$, you must type the index, such as 4 for a 4th root, before selecting $\sqrt[x]{}$.

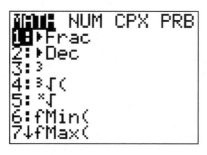

EXAMPLE 16   Using a calculator, find a decimal approximation for $\sqrt[3]{100}$.

Solution: Press $\boxed{\text{MATH}}\,\boxed{4}\,\boxed{1}\,\boxed{0}\,\boxed{0}\,\boxed{)}\,\boxed{\text{ENTER}}$

$$
\begin{array}{l}
\sqrt[3]{100}\\
\qquad\qquad 4.641588834
\end{array}
$$

Answer: 4.642, rounded to three decimal places

EXAMPLE 17   Using a calculator, find a decimal approximation for $\sqrt[4]{20}$.

Solution: Press $\boxed{4}\,\boxed{\text{MATH}}\,\boxed{5}\,\boxed{2}\,\boxed{0}\,\boxed{)}\,\boxed{\text{ENTER}}$

$$
\begin{array}{l}
\sqrt[4]{20}\\
\qquad\qquad 2.114742527
\end{array}
$$

Answer: 2.115, rounded to three decimal places

You should now be able to simplify expressions involving fractional exponents, whether they occur in numerical or in algebraic problems. Note that we are concerned mainly with the ability to rewrite fractional exponents using radicals and are *not* concerned about using a calculator to find decimal approximations.

Before beginning the next unit, you should simplify the following expressions involving fractional exponents.

EXERCISES

Simplify the following without using a calculator:

1. $125^{1/3}$

2. $(-1)^{2/3}$

3. $(-4)^{1/2}$ Careful!

4. $4^{3/2}$

5. $4^{1/2}$

6. $4^{-1/2}$

Rewrite the following expressions using radicals. Simplify whenever possible.

7. $x^{-1/2}$

8. $x^{1/3}$

9. $a^{2/5}$

10. $4^{-3/2}$

11. $(x+1)^{1/2}$

12. $x^{8/3}$

13. $(4x)^{1/2}$

14. $x^{11/2}$

15. $(5x)^{-1/2}$

16. $(18x^3)^{1/2}$

17. $(2x)^{2/3}$

18. $(-64)^{2/3}$

19. Rewrite using a fractional exponent: $\sqrt{7x}$.

20. Rewrite using a fractional exponent: $\sqrt[3]{2x}$.

21. Rewrite using a fractional exponent: $3\sqrt[4]{x}$

22. Rewrite using a fractional exponent: $5\sqrt[3]{y^2}$

23. Rewrite using a fractional exponent: $\dfrac{4}{\sqrt{x}}$

24. Rewrite using a fractional exponent: $\dfrac{1}{7\sqrt[3]{z}}$

C25. Find the decimal approximation correct to three decimal places for $23^{1/5}$.

C26. Find the decimal approximation correct to three decimal places for $\sqrt[4]{17}$.

Solutions to Problems

1. $32^{1/5} = \sqrt[5]{32} = 2$

2. $(-1)^{1/3} = \sqrt[3]{-1} = -1$

3. $4^{5/2} = \left(\sqrt{4}\right)^5 = 2^5 = 32$

4. $64^{2/3} = \left(\sqrt[3]{64}\right)^2 = 4^2 = 16$

5. $81^{3/4} = \left(\sqrt[4]{81}\right)^3 = 3^3 = 27$

6. $(-8)^{2/3} = \left(\sqrt[3]{-8}\right)^2 = (-2)^2 = 4$

- -

UNIT 14

Simplifying Expressions with Fractional Exponents

You have completed all the definitions and laws of exponents necessary to handle any given expression involving exponents. From here on, you need practice, confidence, and a bit of luck! We're going to work on practice in this unit.

The definitions and laws of exponents are restated below for reference:

Positive Integer:	$b^n = \underbrace{b \cdot b \cdot b \cdot \ldots \cdot b}_{n \text{ factors}}$
Zero:	$b^0 = 1$
Negative:	$b^{-n} = \dfrac{1}{b^n}$ and $\dfrac{1}{b^{-n}} = b^n$
Fractional:	$b^{n/d} = \sqrt[d]{b^n} = \left(\sqrt[d]{b}\right)^n$

Laws of Exponents

I.	**Multiplication**	$b^n \cdot b^m = b^{n+m}$	add exponents
II.	**Power of a power**	$(b^n)^m = b^{nm}$	
III.	**Power of a product**	$(bc)^n = b^n c^n$	multiply exponents
IV.	**Power of a fraction**	$\left(\dfrac{a}{b}\right)^n = \dfrac{a^n}{b^n}$	
V.	**Division**	$\dfrac{b^m}{b^n} = b^{m-n}$ or, alternatively, $= \dfrac{1}{b^{n-m}}$	subtract exponents

These problems look tricky. Luckily, fractional exponents obey all the laws of integer exponents.

To simplify expressions involving exponents means to write the final answer without using zero, negative, or fractional exponents. To accomplish this, as in previous units, we **always** will use the same procedure when simplifying. Of course, once again an additional step has been added, this time to include fractional exponents.

My suggested procedure for simplifying an algebraic expression with exponents is:

• If there are parentheses, remove them using the power laws (multiply exponents).

• If there are negative exponents, use the definition to rewrite them as positive exponents.

• If applicable, use the division law (subtract the exponents).

• If applicable, use the multiplication law (add the exponents).

• If fractional exponents remain, use the definition to rewrite as a radical.

It is the same basic pattern we have been using since Unit 7.

EXAMPLE 1

Simplify: $\left(x^{-1/2}\right)^{-2/3}$.

Solution: $\left(x^{-1/2}\right)^{-2/3} = x^{1/3}$ Remove parentheses; multiply exponents.

$$\frac{-1}{2} \cdot \frac{-2}{3} = \frac{1}{3}$$

$$= \sqrt[3]{x} \qquad \text{Rewrite as a radical.}$$

EXAMPLE 2

Simplify: $x^{-1/3} \cdot x^{1/2}$.

Solution: $x^{-1/3} \cdot x^{1/2} = x^{1/6}$ Use the multiplication law; add exponents.

$$\frac{-1}{3} + \frac{1}{2} = \frac{-2+3}{6} = \frac{1}{6}$$

$$= \sqrt[6]{x} \qquad \text{Rewrite as a radical.}$$

EXAMPLE 3

Simplify: $(9x^{-4})^{1/2}$.

Solution: $(9x^{-4})^{1/2} = 9^{1/2} \cdot x^{-2}$ Remove parentheses; multiply exponents.

$$1 \cdot \frac{1}{2} = \frac{1}{2} \text{ and } (-4) \cdot \frac{1}{2} = \frac{\overset{2}{\cancel{-4}}}{1} \cdot \frac{1}{\underset{1}{\cancel{2}}} = -2$$

$$= \frac{9^{1/2}}{x^2}$$ Rewrite with positive exponents.

$$= \frac{3}{x^2}$$ since $9^{1/2} = \sqrt{9} = 3$.

EXAMPLE 4

Simplify: $(x^6 y^{-3})^{-2/3}$.

Solution: $(x^6 y^{-3})^{-2/3} = x^{-4} y^2$ Remove parentheses; multiply exponents.

$$\frac{\overset{2}{\cancel{6}}}{1} \cdot \frac{-2}{\underset{1}{\cancel{3}}} = -4 \text{ and } \frac{\overset{1}{\cancel{-3}}}{1} \cdot \frac{-2}{\underset{1}{\cancel{3}}} = 2$$

$$= \frac{y^2}{x^4}$$ Rewrite with positive exponents.

You try one; then we'll have some more examples.

Problem 1

Simplify: $(x^{-8} y^4)^{-3/2}$.

Solution:

EXAMPLE 5

Simplify: $\left(\dfrac{a^{1/2}b^{2/3}}{c^{1/7}}\right)^{6}$.

Solution: $\left(\dfrac{a^{1/2}b^{2/3}}{c^{1/7}}\right)^{6} = \dfrac{a^{3}b^{4}}{c^{6/7}}$ Remove parentheses; multiply exponents.

$$\frac{1}{\cancel{2}}\cdot\frac{\overset{3}{\cancel{6}}}{1}=3, \quad \frac{2}{\cancel{3}}\cdot\frac{\overset{2}{\cancel{6}}}{1}=4 \text{ and } \frac{1}{7}\cdot\frac{6}{1}=\frac{6}{7}$$

$$= \frac{a^{3}b^{4}}{\sqrt[7]{c^{6}}}$$ Rewrite fractional exponent as a radical.

At this point we're not at all concerned about rationalizing that denominator. Recall from high school days that it was a "no-no" to leave the radical in the denominator.

Quite often expressions you are trying to simplify contain radicals. Begin by using the definition of a fractional exponent to rewrite the radical expression using exponents. This requires adding one final step to our suggested procedure for simplifying algebraic expressions involving exponents.

1. If there are radicals, use the definition to rewrite them as fractional exponents.

2. If there are parentheses, remove them using the power laws (multiply exponents).

3. If there are negative exponents, use the definition to rewrite them as positive exponents.

4. If applicable, use the division law of exponents (subtract exponents).

5. If applicable, use the multiplication law of exponents (add exponents).

6. If fractional exponents remain, use the definition to rewrite them as radicals.

You will be happy to learn that in the majority of the problems encountered in this book, it will not be necessary to use all six steps. Often one or two steps become unnecessary, but the order of the steps **always** remains the same in our examples.

EXAMPLE 6

Simplify: $\sqrt{x}\cdot\sqrt[3]{x}$.

Solution: $\sqrt{x}\cdot\sqrt[3]{x} = x^{1/2}\cdot x^{1/3}$ Rewrite the radicals using fractional exponents.

$\qquad\qquad\quad = x^{5/6}$ Use the multiplication law; add exponents.

$$\frac{1}{2}+\frac{1}{3}=\frac{3+2}{6}=\frac{5}{6}$$

$\qquad\qquad\quad = \sqrt[6]{x^{5}}$ Rewrite the fractional exponent as a radical.

EXAMPLE 7

Simplify: $\dfrac{\sqrt[3]{x^2}}{\sqrt[4]{x}}$.

Solution: $\dfrac{\sqrt[3]{x^2}}{\sqrt[4]{x}} = \dfrac{x^{2/3}}{x^{1/4}}$ Rewrite radicals using fractional exponents.

$= x^{5/12}$ Use the division law; subtract exponents.

$$\frac{2}{3} - \frac{1}{4} = \frac{8-3}{12} = \frac{5}{12}$$

$= \sqrt[12]{x^5}$ Rewrite the fractional exponent as a radical.

EXAMPLE 8

Simplify: $\left[\left(\dfrac{\sqrt{2}}{-3}\right)^{-4}\right]^{-1}$.

Solution: $\left[\left(\dfrac{(2)^{1/2}}{-3}\right)^{-4}\right]^{-1} = \left[\dfrac{(2)^{-2}}{(-3)^{-4}}\right]^{-1} = \dfrac{2^2}{(-3)^4} = \dfrac{4}{81}$

As always, work from the inside out to remove parentheses.

Here are four problems for you.

Problem 2

Simplify: $\left(\sqrt[3]{7}\right)^6$.

Solution:

Problem 3

Simplify: $\left(x^{1/3}\right)^3$.

Solution:

Problem 4

Simplify: $\left(\dfrac{2^0}{8^{1/3}}\right)^{-1}$.

Solution:

Problem 5

Simplify: $\left(2\sqrt[6]{x}\right)^3$.

Solution:

In case you had difficulty with Problems 2–5, let us restate our objective and our procedure.

Objective: To be able to simplify expressions involving exponents. By "simplify" we mean to write the expression without using parentheses or fractional, negative, or zero exponents.

Our procedure: The six steps are listed before Example 6. There is nothing sacred about the order of the steps. As stated before, we always do them in the same order so that you can follow what we are doing in the solutions.

The following example illustrates this procedure one more time.

EXAMPLE 9

Simplify: $\left(\dfrac{\sqrt{x}}{x^2}\right)^{-2}$

Solution: $\left(\dfrac{\sqrt{x}}{x^2}\right)^{-2}$

Step 1. $\left(\dfrac{x^{1/2}}{x^2}\right)^{-2}$

Step 2. $\dfrac{x^{-1}}{x^{-4}}$

Step 3. $\dfrac{x^4}{x}$

Step 4. x^3

If you have made it this far, congratulations. In our opinion, you have just completed the most difficult unit in the book.

You should now be able to simplify most expressions involving exponents, whether they are fractional, negative, zero, or integers, as well as being able to simplify expressions with radicals.

Before beginning the next unit you should simplify the expressions in the exercises. The final answers should be written without fractional, negative, or zero exponents. Also, fractions should be in lowest terms. When you have completed the exercises, check your answers against those given at the back of the book.

EXERCISES

Simplify:

1. $\dfrac{y^{2/3}}{y^{1/3}}$

2. $\left(y^{3/5}\right)^{1/4}$

3. $x^{1/2} \cdot x^{2/5}$

4. $\left(\dfrac{a^4}{c^2}\right)^{1/2}$

5. $\left[\left(\sqrt{4}\right)^{-1}\right]^2$

6. $\left(\sqrt{x}\right)^{1/2}$

7. $\left(8x^2\right)^{1/3}$

8. $\left(\dfrac{2^{-3} \cdot 2^5}{2^{-2}}\right)^3$

9. $\left(\dfrac{x^{1/3}}{x^{2/3}}\right)^3$

10. $x^{1/2} \cdot x^{5/2}$

11. $\left(\sqrt[3]{x^2}\right)^{1/2}$

12. $\dfrac{x^{-7/2} \cdot x^{3/2}}{\sqrt{x} \cdot x^{-3/2}}$

13. $\left(8\sqrt{x}\right)^{-2/3}$

14. $\left(\dfrac{27^{5/3} \cdot 27^{-1/3}}{27^{1/3}}\right)^2$

15. $\left(\dfrac{3x^{-1}}{\sqrt{x}}\right)^2$

Solutions to Problems

1. $\left(x^{-8}y^4\right)^{-3/2} = x^{12}y^{-6} = \dfrac{x^{12}}{y^6}$

   Multiply exponents

   $$\dfrac{\cancel{-8}^{\,-4}}{1} \cdot \dfrac{-3}{\cancel{2}} = 12$$

   $$\dfrac{\cancel{4}^{\,2}}{1} \cdot \dfrac{-3}{\cancel{2}} = -6$$

2. $\left(\sqrt[3]{7}\right)^6 = \left(7^{1/3}\right)^6 = 7^2 = 49$

3. $\left(x^{1/3}\right)^3 = x^1 = x$

4. $\left(\dfrac{2^0}{8^{1/3}}\right)^{-1} = \dfrac{2^0}{8^{-1/3}} = \dfrac{1}{8^{-1/3}} = 8^{1/3} = \sqrt[3]{8} = 2$

5. $\left(2\sqrt[6]{x}\right)^3 = \left(2x^{1/6}\right)^3 = 2^3 x^{1/2} = 8\sqrt{x}$

UNIT 15

Additional Practice with Exponents

Before we leave our study of exponents, we will look at a source of common errors in working with them. These are errors in dealing with the addition and subtraction of terms with exponents.

The five laws of exponents deal only with simplifying products and quotients. In fact, addition and subtraction involving terms with exponents must be simplified using the definitions covered in Units 10 and 11 on fractions.

EXAMPLE 1

Simplify: $(ab)^{-1}$

_ _

Solution: $(ab)^{-1} = a^{-1}b^{-1} = \dfrac{1}{ab}$

_ _

EXAMPLE 2

Simplify: $(a + b)^{-1}$.

_ _

Solution: $(a+b)^{-1} = \dfrac{1}{(a+b)} = \dfrac{1}{a+b}$

_ _

EXAMPLE 3

Simplify: $a^{-1} + b^{-1}$.

Solution: $a^{-1} + b^{-1} = \dfrac{1}{a} \searrow \dfrac{1}{b} = \dfrac{b+a}{ab}$

The arrow is a reminder to add by using the definition from Unit 10.

Be sure you see the difference between Example 1, in which the exponent applies to the product of a and b; Example 2, in which the exponent applies to the sum of a and b; and Example 3, in which the exponents apply to the two separate terms.

EXAMPLE 4

Simplify: $a + b^{-1}$.

Solution: $a + b^{-1} = a + \dfrac{1}{b} = \dfrac{a}{1} \searrow \dfrac{1}{b} = \dfrac{ab+1}{b}$

Try this problem.

Problem 1

Simplify: $3^{-1} + x^{-1}$.

Solution:

By now a problem like the following should be easy for you.

Problem 2

Simplify: $\dfrac{a^{-1}b^{-1}}{c^{-1}d^{-1}}$.

Solution:

Here's another example.

EXAMPLE 5

Simplify: $\dfrac{a^{-1}+b^{-1}}{c^{-1}+d^{-1}}$.

Solution: $\dfrac{a^{-1}+b^{-1}}{c^{-1}+d^{-1}} = \dfrac{\dfrac{1}{a}+\dfrac{1}{b}}{\dfrac{1}{c}+\dfrac{1}{d}}$, which is a complex fraction

$$= \left(\dfrac{1}{a}+\dfrac{1}{b}\right) \div \left(\dfrac{1}{c}+\dfrac{1}{d}\right)$$

$$= \dfrac{b+a}{ab} \div \dfrac{d+c}{cd}$$

$$= \dfrac{a+b}{ab} \cdot \dfrac{cd}{c+d}$$

$$= \dfrac{cd(a+b)}{ab(c+d)}$$

Now you try.

Problem 3

Simplify: $\dfrac{1}{a^{-1}+b^{-1}}$.

Solution:

Before proceeding, be sure you understand how Problem 2, Example 5, and Problem 3 differ.

Now consider Examples 6 and 7.

EXAMPLE 6

Simplify: $\dfrac{1}{x^{-1}+x}$.

Solution: $\dfrac{1}{x^{-1}+x}=\dfrac{1}{\dfrac{1}{x}+x}$, which is a complex fraction

$$= 1 \div \left(\frac{1}{x} + x \right)$$

$$= 1 \div \left(\frac{1}{x} + \frac{x}{1} \right)$$

$$= 1 \div \frac{1+x^2}{x}$$

$$= 1 \cdot \frac{x}{1+x^2}$$

$$= \frac{x}{1+x^2}$$

Let me repeat: the laws of exponents deal only with multiplication and division. There are no laws dealing with powers of sums and differences!

EXAMPLE 7

Simplify: $(x^2 + a)^{-3}$

Solution: $(x^2 + a)^{-3} = \dfrac{1}{(x^2 + a)^3}$ and that is it.

You should now be able to simplify expressions in which terms with exponents are added and subtracted. Remember that, since the laws of exponents deal only with multiplication and division, you must call upon the definitions for addition and subtraction of fractions and the techniques for simplifying complex fractions.

Before beginning the next unit, simplify the expressions in the exercises, writing them as either monomials or fractions in lowest terms, with positive exponents only. None of the denominators is zero.

EXERCISES

Simplify:

1. $(x^2 + 1)^{-2}$

2. $x^{-1} + 2^{-2}$

3. $x^{-1} - 1$

4. $\dfrac{2}{x^{-1} + 3y^{-1}}$

5. $3^{-1} + 3^{-2}$

6. $x^{-1} + 2y^{-1}$

7. $(ab^{-1})^{-2}$

8. $5(x + y)^{-1}$

9. $\dfrac{x^{-1}}{(2x - 3)^{-2}}$

10. $3x^{-2} + y$

11. $\dfrac{a - a^{-1}}{a + a^{-1}}$

12. $\dfrac{3^{-1} + 2^{-1}}{3^{-1} - 2^{-1}}$

13. $\dfrac{a + b^{-1}}{ab}$

14. $\dfrac{3ab}{a^{-1} + b}$

Solutions to Problems

1. $3^{-1} + x^{-1} = \dfrac{1}{3} + \dfrac{1}{x} = \dfrac{x + 3}{3x}$

2. $\dfrac{a^{-1}b^{-1}}{c^{-1}d^{-1}} = \dfrac{c^1 d^1}{a^1 b^1} = \dfrac{cd}{ab}$

3. $\dfrac{1}{a^{-1} + b^{-1}} = \dfrac{1}{\dfrac{1}{a} + \dfrac{1}{b}}$

$= 1 \div \left(\dfrac{1}{a} + \dfrac{1}{b}\right)$

$= 1 \div \left(\dfrac{b + a}{ab}\right)$

$= \dfrac{1}{1} \cdot \dfrac{ab}{b + a}$

$= \dfrac{ab}{b + a}$

UNIT 16

Multiplication of Monomials and Polynomials

In this unit we will learn how to multiply binomials and other polynomials. We will discuss some methods that should help you perform these operations and avoid common errors.

Recall that expressions with one term are called **monomials** and that expressions with more than one term (symbols or groups of symbols separated by a plus or minus sign) are called **polynomials**. Polynomials are sometimes classified according to the number of terms; for example, a **binomial** has two terms and a **trinomial** has three terms.

MULTIPLICATION BY A MONOMIAL

We will first consider multiplication by a monomial.

Rule: **To multiply two monomials, multiply their numerical coefficients and find the product of the variable factors according to the laws of exponents.**

Recall that since we are multiplying (multiplication law of exponents), the exponents are added.

EXAMPLE 1 $(2x^3)(3x^5) = 2 \cdot 3 \cdot x^3 \cdot x^5 = 6x^8$

EXAMPLE 2 $(2xy^3)(7x^4y^5) = 2 \cdot 7 \cdot x \cdot x^4 \cdot y^3 \cdot y^5 = 14x^5y^8$

EXAMPLE 3 $(3\pi ab)(2\pi a^2bc^3) = 3 \cdot 2 \cdot \pi \cdot \pi \cdot a \cdot a^2 \cdot b \cdot b \cdot c^3 = 6\pi^2a^3b^2c^3$

The above rule, combined with the distributive property $a(b + c) = ab + ac$, extends in the natural way to multiplication of a polynomial by a monomial.

EXAMPLE 4 $2x^3(3x^5 + 2xy^3) = 6x^8 + 4x^4y^3$

EXAMPLE 5 $2xy^3(3x^5 + 7x^4y + 1) = 6x^6y^3 + 14x^5y^4 + 2xy^3$

EXAMPLE 6 $-3a^2b(2a^4b^2 - 6a^3b) = -6a^6b^3 + 18a^5b^2$

Here are three problems for you.

Problem 1 $3x^3(5x^2 - 2) =$

Problem 2 $3xy^2(2xy + 3x^2y^3) =$

Problem 3 $2xyz(x + y + 2z + 3) =$

MULTIPLICATION OF TWO POLYNOMIALS

Now let's consider multiplication of two polynomials. This is the most frequently encountered problem, and ideally you should be able to do many of these multiplications mentally.

Consider, for example, $(x + 3)(x + 4)$.

Basically, three methods are commonly used to find the product of two binomials, $(x + 3)(x + 4)$.

Method 1

$$
\begin{array}{r}
x + 3 \\
x + 4 \\
\hline
x^2 + 3x \\
4x + 12 \\
\hline
x^2 + 7x + 12
\end{array}
$$

This way of multiplying polynomials is not very efficient. We generally discourage its use, primarily because it necessitates rewriting the problem.

Method 2 (FOIL)

FOIL   for   F = first $= x \cdot x$
O = outer $= 4 \cdot x$
I = inner $= 3 \cdot x$
L = last $= 3 \cdot 4$

F   L = F   O   I   L
$(x + 3)(x + 4) = x^2 + 4x + 3x + 12 = x^2 + 7x + 12$
I
O

The advantage of the FOIL method is that for simple binomials the multiplication can be performed mentally. However, it can be used only when multiplying two binomials together, *not* when multiplying a binomial by some other polynomial.

Method 3 (distributive property)

$$(x + 3)(x + 4) = x(x + 4) + 3(x + 4)$$
$$= x^2 + 4x + 3x + 12$$
$$= x^2 + 7x + 12$$

This method is similar to removing parentheses and works for multiplying *any* type of polynomial. We think it is the most useful method and the one we will use in all future examples.

EXAMPLE 7

Multiply: $(x - 7)(x + 2)$.

_____

Solution: $(x - 7)(x + 2) = x(x + 2) - 7(x + 2)$
$$= x^2 + 2x - 7x - 14$$
$$= x^2 - 5x - 14$$

_____

Problem 4

Multiply: $(2x + 5)(3x - 2)$.

_____

Solution:

_____

After this example, it will be your turn again.

EXAMPLE 8

Multiply: $(x + a)(2x - b)$.

Solution: $(x + a)(2x - b) = x(2x - b) + a(2x - b)$
$$= 2x^2 - bx + 2ax - ab$$

Problem 5

Multiply: $(3x + 1)(x - 5)$.

Solution:

Problem 6

Multiply: $(3x - 5)(2x - 1)$.

Solution:

The next example illustrates the fact that by using the distributive property we can multiply any polynomial by a binomial.

EXAMPLE 9

Multiply: $(x + 4)(x^3 - x^2 + 3x - 1)$.

Solution: $(x + 4)(x^3 - x^2 + 3x - 1) = x(x^3 - x^2 + 3x - 1) + 4(x^3 - x^2 + 3x - 1)$
$$= x^4 - x^3 + 3x^2 - x + 4x^3 - 4x^2 + 12x - 4$$
$$= x^4 + 3x^3 - x^2 + 11x - 4$$

Note that the answer is written in descending order of the exponents, ending with the constant term.

EXAMPLE 10

Multiply: $(x + 2)(x^3 - 1)$.

Solution: $(x + 2)(x^3 - 1) = x^4 - x + 2x^3 - 2$
$$= x^4 + 2x^3 - x - 2$$

Before completing the unit, let's look at several more examples that illustrate a "special product"—a binomial squared.

EXAMPLE 11

Multiply: $(x + 3)^2$.

Solution: $(x + 3)^2 = (x + 3)(x + 3)$
$$= x(x + 3) + 3(x + 3)$$
$$= x^2 + 3x + 3x + 9$$
$$= x^2 + 6x + 9$$

Note that the answer has the first and last terms squared but there is also **a middle term that is twice the product of the two terms** of the binomial.

$$(a + b)^2 = (a + b)(a + b)$$
$$= a(a + b) + b(a + b)$$
$$= a^2 + ab + ab + b^2$$
$$= a^2 + 2ab + b^2$$

This problem and Example 11 illustrate the following statement:

When a binomial is squared, the result will be the first term squared plus twice the product of the two terms plus the last term squared.

$$(a + b)^2 = a^2 + 2ab + b^2$$

first term   twice the   last term
squared   product of   squared
the terms

EXAMPLE 12

Multiply: $(x + 7)^2$.

Solution:
$$(x + 7)^2 = x^2 + 2(7x) + 7^2$$
$$= x^2 + 14x + 49$$

EXAMPLE 13

Multiply: $(x - 5)^2$.

Solution:
$$(x - 5)^2 = x^2 + 2(-5x) + 25$$
$$= x^2 - 10x + 25$$

EXAMPLE 14

Multiply: $x(x + 4)^2$.

Solution: As always, remove parentheses first.
$$x(x + 4)^2 = x(x^2 + 2(4x) + 16)$$
$$= x(x^2 + 8x + 16)$$
$$= x^3 + 8x^2 + 16x$$

EXAMPLE 15

Multiply: $2x^3(x - 2)^2$.

Solution:
$$2x^3(x - 2)^2 = 2x^3(x^2 + 2(-2x) + 4)$$
$$= 2x^3(x^2 - 4x + 4)$$
$$= 2x^5 - 8x^4 + 8x^3$$

You should now be able to multiply polynomials. Again we suggest using the distributive property, which is similar to removing parentheses and will work for all cases.

Before beginning the next unit you should try to solve the following exercises.

EXERCISES

In each case, perform the indicated multiplication.

1. $2cx^2(5c^2 - c - 3x)$
2. $(x + 4)(x + 5)$
3. $(x - 7)(x - 2)$
4. $(x - 1)(x - 5)$
5. $(x + 2)(x - 3)$
6. $(a + 5)^2$
7. $(x + 2)(x - 2)$
8. $(x - 1)^2$
9. $(x - 4)(x + 3)$
10. $(2x + 1)(x + 1)$
11. $(2x - 5)(x + 4)$
12. $(3x - 2)(x + 7)$
13. $(5x + 1)(x + 2)$
14. $(3x + 1)(3x - 1)$
15. $(2x + 3)(4x + 1)$
16. $(5x - 1)(x + 2)$
17. $(2x + 3)^2$
18. $(4x - 3)^2$
19. $(3a + b)(2a - b)$
20. $(x + y)(x - y)$
21. $(3x + 1)(2x - 5)$
22. $x(x - 4)^2$
23. $3x^2(2x + 5)^2$
24. $(x - 2)(x^3 - 4x^2 + 7x - 1)$
25. $(x^2 + 1)(x^2 - 3)$
26. $(x + 2y)(x - 3y)$
27. $(2a - 1)(3 - a)$
28. $(x^2 - 3x + 1)(x^3 - 2x)$
29. $(x^2 + 5)(x - 3)$
30. $(5a - 3b)(-2a + 6b)$

Solutions to Problems

1. $3x^3(5x^2 - 2) = 15x^5 - 6x^3$

2. $3xy^2(2xy + 3x^2y^3) = 6x^2y^3 + 9x^3y^5$

3. $2xyz(x + y + 2z + 3) = 2x^2yz + 2xy^2z + 4xyz^2 + 6xyz$

4. $(2x + 5)(3x - 2) = 2x(3x - 2) + 5(3x - 2)$

$$= 6x^2 - 4x + 15x - 10$$

$$= 6x^2 + 11x - 10$$

5. $(3x + 1)(x - 5) = 3x(x - 5) + 1(x - 5)$

$$= 3x^2 - 15x + x - 5$$

$$= 3x^2 - 14x - 5$$

6. $(3x - 5)(2x - 1) = 3x(2x - 1) - 5(2x - 1)$

$$= 6x^2 - 3x - 10x + 5 \quad \text{Note the sign on } +5$$

$$= 6x^2 - 13x + 5$$

UNIT 17

Division of Polynomials

In this unit we will examine the procedure for dividing polynomials. Although this kind of division does not occur very often, it is occasionally necessary to perform the operation in the course of solving a problem. The process outlined here will help on such occasions.

TWO IMPORTANT DEFINITIONS

Before studying the procedure for dividing polynomials, let's look at two definitions.

> Definition: A **polynomial** in x is said to be in **standard form** if:
> 1. All parentheses are removed.
> 2. Like terms are combined.
> 3. The terms are arranged in order of descending powers of x.

> Definition: The **degree of polynomial** in x is the greatest power of x.

EXAMPLE 1

Write $x + 2x(x^2 - 5)$ in standard form and find its degree.

Solution: $x + 2x(x^2 - 5) = x + 2x(x^2 - 5)$

$\qquad\qquad\qquad = x + 2x^3 - 10x$

$\qquad\qquad\qquad = 2x^3 - 9x$

Standard form: $2x^{③} - 9x$

Degree: 3

PROCEDURE FOR DIVIDING POLYNOMIALS

Now let's look at the procedure for dividing polynomials. Consider this problem:

Divide: $(5x^2 - 3x + 1)$ by $(x - 2)$.

The basic procedure parallels the long-division process in arithmetic.

Step 1:   Arrange both dividend and divisor in standard form and set up as a long-division problem, writing any missing terms with 0 for a coefficient.

$$x - 2\overline{)5x^2 - 3x + 1}$$

Step 2:   Divide the first term of the divisor into the first term of the dividend. (Recall that, since we are dividing, the exponents are subtracted.)

$$\dfrac{5x^2}{x} = 5x$$

$$\begin{array}{r} 5x \\ x - 2\overline{)5x^2 - 3x + 1} \end{array}$$

Step 3:   Multiply each term of the divisor by the first term of the quotient.

$$5x(x - 2) = 5x^2 - 10x$$

$$\begin{array}{r} 5x \\ x - 2\overline{)5x^2 - 3x + 1} \\ 5x^2 - 10x \end{array}$$

Step 4:   Subtract like terms by changing to addition of the opposite and then combining, and bring down the next term from the dividend.

$$\begin{array}{c} 5x^2 - 3x + 1 \\ -(5x^2 - 10x) \end{array} \rightarrow \begin{array}{c} 5x^2 - 3x + 1 \\ -5x^2 + 10x \end{array}$$

$$\begin{array}{r} 5x \\ x - 2\overline{)5x^2 - 3x + 1} \\ -5x^2 + 10x \\ + 7x + 1 \end{array}$$

Step 5:   Repeat steps 2, 3, and 4, using the new remainder, $7x + 1$, as the dividend.

$$\dfrac{7x}{x} = 7$$

$$\begin{array}{r} 5x + 7 \\ x - 2\overline{)5x^2 - 3x + 1} \\ -5x^2 + 10x \\ 7x + 1 \\ -7x + 14 \\ \hline 15 \end{array}$$

Step 6:   Continue repeating steps 2, 3, and 4 until the degree of the remainder is less than the degree of the divisor.

$$7(x - 2) = 7x - 14 \qquad \begin{array}{c} 7x + 1 \\ -(7x - 14) \end{array} \rightarrow \begin{array}{c} 7x + 1 \\ -7x + 14 \end{array}$$

Step 7: Any non-zero remainder is written as a fraction with the divisor as the denominator.

$$\frac{15}{x-2}$$

Therefore

$(5x^2 - 3x + 1)$ divided by $(x - 2)$ equals $5x + 7 + \dfrac{15}{x-2}.$

Let us now try condensing some of the explanation.

EXAMPLE 2

Divide: $(5x^2 + 7x - 1)$ by $(x + 3)$.

Solution: Step 2. Divide first terms:

$$\frac{5x^2}{x} = 5x.$$

$$\begin{array}{r} 5x \\ x + 3 \overline{)\ 5x^2 + 7x - 1} \\ \underline{-5x^2 - 15x} \\ -8x - 1 \end{array}$$

Step 3. Multiply:

$$5x(x + 3) = 5x^2 + 15x.$$

Step 4. Add the opposite; bring down $- 1$.

Repeat steps 2, 3, and 4, using $-8x - 1$.

Step 2. Divide first terms:

$$\frac{-8x}{x} = -8.$$

$$\begin{array}{r} 5x - 8 \\ x + 3 \overline{)\ 5x^2 + 7x - 1} \\ \underline{-5x^2 - 15x} \\ -8x - 1 \\ \underline{+8x + 24} \\ 23 \end{array}$$

Step 3. Multiply:

$$-8(x + 3) = -8x - 24.$$

Step 4. Add the opposite.

The division process is finished because the degree of the remainder is less than the degree of the divisor.

Write the remainder as a fraction.

$$\begin{array}{r} 5x - 8 + \dfrac{23}{x + 3} \\ x + 3 \overline{)\ 5x^2 + 7x - 1} \\ \underline{-5x^2 - 15x} \\ -8x - 1 \\ \underline{+8x + 24} \\ +23 \end{array}$$

Therefore:

$$(5x^2 + 7x - 1) \div (x + 3) = 5x - 8 + \frac{23}{x + 3}.$$

Here are two more examples. Example 3 is given in detail, with explanation.

EXAMPLE 3

Divide: $(6x^2 + 23x + 20)$ by $(2x + 5)$

Solution: Step 1.   Set up problem.

Step 2.   Divide first terms:

$$\frac{6x^2}{2x} = 3x.$$

Step 3.   Multiply:

$3x(2x + 5)$.

Step 4.   Add the opposite and bring down

the next term of 20.

$$\begin{array}{r} 3x + 4 \\ 2x + 5 \overline{)\, 6x^2 + 23x + 20} \\ \underline{-6x^2 - 15x} \\ 8x + 20 \\ \underline{-8x - 20} \\ 0 \end{array}$$

Repeat steps 2, 3, and 4 using $8x + 20$.

Step 2.   Divide first terms:

$$\frac{8x}{2x} = 4.$$

Step 3.   Multiply:

$4(2x + 5)$.

Step 4.   Add the opposite.

The division process is finished, and there is no remainder.

Answer: $3x + 4$

EXAMPLE 4

Divide: $(x^2 - x - 15)$ by $(x + 1)$

Solution:

$$\begin{array}{r} x - 2 \\ x + 1 \overline{)\, x^2 - x - 15} \\ \underline{-x^2 - x} \\ -2x - 15 \\ \underline{+2x +\ 2} \\ -13 \end{array}$$

Answer: $\quad x - 2 + \dfrac{-13}{x + 1}.$

You try this one.

Problem 1

Divide: $\quad (3x^2 + 4x + 1)$ by $(3x + 1)$.

Solution:

EXAMPLE 5

Divide: $\quad (6x + 5)$ by $(2x - 1)$

Solution:

$$\begin{array}{r} 3 \\ 2x - 1 \overline{)\, 6x + 5} \\ \underline{6x - 3} \\ +8 \end{array}$$

Answer: $\quad 3 + \dfrac{8}{2x - 1}$

Look carefully at Examples 6–7.

EXAMPLE 6

Divide: $(4x^2 - 7)$ by $(3 + x)$.

Solution:

$$
\begin{array}{r}
4x - 12 \\
x + 3 \overline{\smash{\big)}\ 4x^2 + 0x - 7} \\
\underline{-4x^2 - 12x} \\
-12x - 7 \\
\underline{+12x + 36} \\
29
\end{array}
$$

Answer: $4x - 12 + \dfrac{29}{x+3}$

> Note: The divisor had to be put in standard form, and a space had to be left for the missing x-term in the dividend. We inserted a "$0x$" in the space to avoid confusion when we subtracted.

EXAMPLE 7

Divide: $(x^3 - 8)$ by $(x - 2)$

Solution:

$$
\begin{array}{r}
x^2 + 2x + 4 \\
x - 2 \overline{\smash{\big)}\ x^3 + 0x^2 + 0x - 8} \\
\underline{-x^3 + 2x^2} \\
+2x^2 + 0x \\
\underline{-2x^2 + 4x} \\
+4x - 8 \\
\underline{-4x + 8} \\
0
\end{array}
$$

Answer: $x^2 + 2x + 4$

You should now be able to divide a polynomial by a binomial. Remember that the procedure is identical to the long-division process in arithmetic.

Recall that, in short, the division procedure is:

1. **Set up** as long division.
2. **Divide** first terms.
3. **Multiply** the quotient times the divisor.
4. **Subtract** like terms by adding the opposite; **bring down** the next term.
5. **Repeat**, using the new remainder as the dividend.
6. **Continue** until the degree of the remainder is less than the degree of the divisor.

Now try the following problems.

EXERCISES

Divide:

1. $(x^2 + 5x + 6) \div (x + 2)$
2. $(12x + 1) \div (x - 4)$
3. $(x^2 + 8x + 16) \div (x + 5)$
4. $(10x^2 - 13x + 1) \div (5x + 1)$
5. $(3x^2 - 10x - 8) \div (3x + 2)$
6. $(6x^2 - 19x + 10) \div (2x - 1)$
7. $(4x^2 - 49) \div (2x - 7)$
8. $(5x^2 - 1) \div (x - 1)$
9. $(x^3 + 1) \div (x + 1)$
10. $(x^3 + 2) \div (x + 2)$

Solutions to Problem

1.
$$
\begin{array}{r}
x + 1 \\
3x + 1 \overline{)\ 3x^2 + 4x + 1} \\
\underline{-3x^2 - x} \\
3x + 1 \\
\underline{-3x - 1} \\
0
\end{array}
$$

UNIT 18

Factoring Polynomials

In this unit we will discuss **factoring** of polynomials. Simply stated, factoring involves finding factors whose products are equal to the original polynomial. We will look first at the simplest case, that of finding a common monomial factor. Then we will consider the factoring of trinomials.

FACTORING OUT A COMMON MONOMIAL FACTOR

First, recall the distributive law: $a(b + c) = ab + ac$.

multiplication $\longrightarrow$ $\longleftarrow$ factoring

Factoring out a common monomial factor is the application of the distributive law read from right to left.

A common monomial factor is a single expression that is a factor of each term of the polynomial.

Factoring out the common monomial factor is, in effect, taking out in front of the parentheses, *as much as possible*, what is common to all terms of the polynomial. It is just the reverse of removing parentheses. You already did this on a limited basis when you solved literal equations in Unit 5.

The test of factoring is always whether you can multiply the factors together and obtain the original polynomial.

EXAMPLE 1 $\qquad$ $6x^2y + 5x^2z = x^2(6y + 5z)$

EXAMPLE 2 $\qquad$ $2ay + 2 = 2(ay + 1)$

EXAMPLE 3 $\qquad$ $2xy - xy^2 + 3x^2y = xy(2 - y + 3x)$

EXAMPLE 4 $\qquad$ $8x^2 - 12x = 4x(2x - 3)$

EXAMPLE 5 $\qquad$ $12a^3 - 18a^2 + 3a = 3a(4a^2 - 6a + 1)$

Try it yourself. Express the following in factored form.

Problem 1 $\qquad$ $2ax^2 - 6ax + 10a$

Problem 2 $\qquad$ $x^4 - 5x^3 + 8x^2$

Problem 3 $\qquad$ $6x^3 - 12x^2 + 3x$

Whenever you are factoring an expression, always first look to see whether there is a common monomial factor! Often the common monomial factor is overlooked, and the resulting problem is more difficult to handle.

FACTORING TRINOMIALS

Now let us look at factoring quadratic trinomials, polynomials of the form $ax^2 + bx + c$. In this unit, we will limit ourselves to the case where $a = 1$ and b and c are non-zero integers. The special case where $b = 0$ is covered in Unit 19, and the harder case where $a \ne 1$ will be done in Unit 20.

> Rule: $x^2 + bx + c = (x + m)(x + n)$ where $mn = c$
> and $m + n = b$

In words, this rule says that to factor a trinomial of the form $x^2 + bx + c$ where b and c are integers, we need to find two integers, m and n, that multiply to c and add to b. If such integers can be found, then $x^2 + bx + c$ factors into $(x + m)(x + n)$.

Finding the integers m and n is usually done by trial and error. First look for pairs of factors of c. If you need to, make a written list of them. Then look for the pair that adds to b. If you know your multiplication and addition facts well, you will often be able to do this in your head.

Remembering the rules for operations with signed numbers will help in finding the necessary integers m and n.

1. If c is positive, then m and n will both have the *same sign*. The signs will be the same as the sign of b.

2. If c is negative, m and n will have *opposite signs*. The sign of the number with the larger absolute value will be the same as the sign of b.

Keep in mind two things. First, not all quadratic trinomials can be factored. If no integers exist that multiply to c and add to b, then the polynomial is said to be **prime**. Second, if two integers do exist that satisfy the necessary conditions, they will be unique. If a quadratic trinomial can be factored at all, it can be factored in only one way.

EXAMPLE 6

Factor: $x^2 + 7x + 12$.

───

Solution: In this problem, $b = 7$ and $c = 12$.

We need two integers where 1) the product is 12 $mn = 12$

and 2) the sum is 7 $m + n = 7$

Since c is positive, we need m and n to have the same sign. Since b is positive, both signs will be positive.

Using only positive factors, 12 can be factored into $1 \cdot 12$, $2 \cdot 6$ or $3 \cdot 4$. Of those choices, only 3 and 4 add up to 7. Therefore:

$$x^2 + 7x + 12 = (x + 3)(x + 4)$$

Note that since multiplication is commutative, this could also be written $(x + 4)$ $(x + 3)$.

───

EXAMPLE 7

Factor: $x^2 - 9x + 14$

───

Solution: In this problem, $b = -9$ and $c = 14$.

We need two integers where 1) the product is 14 $mn = 14$

and 2) the sum is -9 $m + n = -9$

Since c is positive, we need m and n to have the same sign. Since b is negative, both signs will be negative.

Using only negative factors, 14 can be factored into $-1 \cdot -14$, or $-2 \cdot -7$. Of those choices, only -2 and -7 add up to -9. Therefore:

$$x^2 - 9x + 14 = (x - 2)(x - 7)$$

───

EXAMPLE 8

Factor: $x^2 - 3x - 10$

Solution: We have $b = -3$ and $c = -10$.

We need two integers where 1) the product is -10 $mn = -10$

and 2) the sum is -3 $m + n = -3$

Since c is negative, we need m and n to have opposite signs. Since b is negative, we need the factor with the larger absolute value to be negative. With these conditions, the possible factors of -10 are $1 \cdot -10$ and $2 \cdot -5$. Of those choices, 2 and -5 have the desired sum of -3. Therefore:

$$x^2 - 3x - 10 = (x + 2)(x + (-5))$$
$$= (x + 2)(x - 5)$$

EXAMPLE 9

Factor: $x^2 + x - 20$

Solution: Remember that even though it is not written, it is understood that $b = 1$; $c = -20$.

We need two integers where 1) the product is -20 $mn = -20$

and 2) the sum is 1 $m + n = 1$

Since c is negative, we need m and n to have opposite signs. Since b is positive, we need the factor with the larger absolute value to be positive. Possible factors of -20 are $-1 \cdot 20$, $-2 \cdot 10$ and $-4 \cdot 5$. Of those, -4 and 5 have the desired sum of 1. Therefore:

$$x^2 + x - 20 = (x - 4)(x + 5)$$

EXAMPLE 10

Factor: $x^2 + 7x + 1$

Solution: We need two positive integers whose product is 1 and whose sum is 7. No such integers exist. Therefore:

$x^2 + 7x + 1$ is prime. It cannot be factored with integers.

EXAMPLE 11 $\quad x^2 + 10x + 9 = (x + 1)(x + 9)$

EXAMPLE 12 $\quad x^2 - 10x + 24 = (x - 4)(x - 6)$

EXAMPLE 13 $\quad x^2 - 10x - 24 = (x + 2)(x - 12)$

EXAMPLE 14 $\quad x^2 + 2x - 35 = (x - 5)(x + 7)$

EXAMPLE 15 $\quad x^2 + 2x + 35$ is prime

EXAMPLE 16 $\quad x^2 - 17x - 60 = (x + 3)(x - 20)$

Factoring is often considered one of the more difficult parts of algebra. Take the opportunity to practice on the following problems.

Problem 4 $\quad x^2 + 5x + 4$

Problem 5 $\quad x^2 - 11x + 24$

Problem 6 $\quad x^2 + 2x - 15$

Problem 7 $\quad x^2 - 6x - 16$

Problem 8 $\quad x^2 - 3x - 54$

Problem 9 $\quad x^2 + 9x + 12$

Problem 10 $\quad x^2 - 12x + 36$

Problem 11 $\quad x^2 + 4x - 21$

FACTORING COMPLETELY

In this unit, we have looked at two techniques for factoring: factoring out a common monomial and factoring quadratic trinomials with $a = 1$. Sometimes expressions need both techniques in order to be factored completely. To factor a polynomial completely:

1. First, factor out common monomial factors, if any.

2. Then, factor the remaining polynomial, if possible.

In the next two units, we will learn some more techniques applicable to the second step. For now, here are a couple of examples that show how to factor completely with the methods we know so far.

EXAMPLE 17

Factor: $\quad 3x^2 + 12x - 36$

Solution: $\qquad 3x^2 + 12x - 36$

$\qquad$ Step 1: $\quad 3(x^2 + 4x - 12)$

$\qquad$ Step 2: $\quad 3(x - 2)(x + 6)$

There are a couple of common errors you should be aware of. Be careful not to lose the common factor while factoring the trinomial. The correct answer to the previous example is $3(x - 2)(x + 6)$, not just $(x - 2)(x + 6)$. Also, do not try to do both steps at the same time. Students who don't write out the first step and try to go straight to the answer in step 2 almost always make mistakes.

EXAMPLE 18

Factor: $5x^4 - 50x^3 + 80x^2$

Solution: $5x^4 - 50x^3 + 80x^2$

Step 1: $5x^2(x^2 - 10x + 16)$

Step 2: $5x^2(x - 2)(x - 8)$

Problem 12 $2x^2 + 4x - 30$
Problem 13 $x^3 + 11x^2 + 28x$
Problem 14 $2ax^2 - 16ax + 30a$

ALGEBRA AND THE CALCULATOR (Optional)

You can use the table feature of a graphing calculator to help you factor trinomials.

EXAMPLE 19

Factor: $x^2 - 10x - 24$

Solution: In this problem, $b = -10$ and $c = -24$.

We need two integers where 1) the product is -24 $mn = -24$

and 2) the sum is -10 $m + n = -10$

There are lots of choices for numbers that multiply to -24. To use a graphing calculator to help, type Y1 = -24/X using these key presses, $\boxed{\text{Y=}}$ $\boxed{\text{CLEAR}}$ $\boxed{(-)}$ $\boxed{2}$ $\boxed{4}$ $\boxed{\div}$ $\boxed{\text{X,T,θ,n}}$ $\boxed{\text{ENTER}}$. Now look in the table, $\boxed{\text{2nd}}$ $\boxed{\text{GRAPH}}$. The pairs of values shown in the table are numbers that multiply to -24. You want a pair that also adds to -10. Only look at integer values in the table; skip over any decimal values or Error.

```
Plot1 Plot2 Plot3      X  | Y1 |
\Y1B-24/X          -2    | 12 |
\Y2=               -1    | 24 |
\Y3=                0    | ERROR |
\Y4=                1    | -24 |
\Y5=                2    | -12 |
\Y6=                3    | -8 |
\Y7=                4    | -6 |
                   X=2
```

Only 2 and –12 add up to –10. Therefore:

$$x^2 - 10x - 24 = (x - 12)(x + 2)$$

EXAMPLE 20

Factor: $x^2 + 22x + 72$

Solution: In this problem, $b = 22$ and $c = 72$.

We need two integers where 1) the product is 72 $mn = 72$

and 2) the sum is 22 $m + n = 22$

You probably know numbers that multiply to 72, but they may not add to 22. To use a graphing calculator, type Y1 = 72/X. Now look in the table. You want a pair that adds to 22. Since the values for both b and c are positive, look only at positive numbers.

```
Plot1 Plot2 Plot3      X  | Y1 |
\Y1B72/X            1    | 72 |
\Y2=               2    | 36 |
\Y3=               3    | 24 |
\Y4=               4    | 18 |
\Y5=               5    | 14.4 |
\Y6=               6    | 12 |
\Y7=               7    | 10.286 |
                   X=4
```

Only 4 and 18 add up to 22. Therefore:

$$x^2 + 22x + 72 = (x + 4)(x + 18)$$

Before beginning the next unit, try factoring the following polynomials. Remember to always first look to see whether there is a common monomial factor that can be factored out before attempting to factor the polynomial into binomial factors.

EXERCISES

Factor:

1. $x^2 + 2x - 3$

2. $x^2 - 15x + 56$

3. $x^2 + x$

4. $3x^2y - 12xy^2$

5. $3bx^2 + 27b^2$

6. $x^2 - x - 6$

7. $x^2 + 5x + 6$

8. $x^2 - 7x + 12$

9. $x^2 - x + 5$

10. $x^2 - 2x - 8$

11. $x^2 + 5x + 3$

12. $2x^3 + 2x^2 + 22x$

13. $5x^2 - 5x - 5$

14. $x^2 + 3x - 10$

15. $x^2 + 7x - 30$

16. $x^2 + 7x + 6$

17. $x^2 + 2x + 1$

18. $x^2 - 6x + 9$

19. $x^2 - x - 56$

20. $x^2 + 4x - 45$

21. $x^2 + 16x + 64$

22. $x^2 - 13x + 40$

23. $x^2 + 7x - 18$

24. $x^3 + x^2 + 5x$

25. $x^2 - 10x + 21$

26. $x^2 - 7x - 18$

27. $7x^2 - 14x + 7$

28. $2x^3 - 6x^2 - 36x$

Solutions to Problems

1. $2ax^2 - 6ax + 10a = 2a(a^2 - 3x + 5)$

2. $x^4 - 5x^3 + 8x^2 = x^2(x^2 - 5x + 8)$

3. $6x^3 - 12x^2 + 3x = 3x(2x^2 - 4x + 1)$

4. $x^2 + 5x + 4 = (x + 1)(x + 4)$

5. $x^2 - 11x + 24 = (x - 3)(x - 8)$

6. $x^2 + 2x - 15 = (x - 3)(x + 5)$

7. $x^2 - 6x - 16 = (x + 2)(x - 8)$

8. $x^2 - 3x - 54 = (x + 6)(x - 9)$

9. $x^2 + 9x + 12$ is prime

10. $x^2 - 12x + 36 = (x - 6)(x - 6)$

11. $x^2 + 4x - 21 = (x - 3)(x + 7)$

12. $2x^2 + 4x - 30 = 2(x^2 + 2x - 15) = 2(x - 3)(x + 5)$

13. $x^3 + 11x^2 + 28x = x(x^2 + 11x + 28) = x(x + 4)(x + 7)$

14. $2ax^2 - 16ax + 30a = 2a(x^2 - 8x + 15) = 2a(x - 3)(x - 5)$

UNIT 19

Factoring a Special Binomial

In this unit we introduce a frequently occurring formula that will aid you in factoring. Also, we consider a suggested procedure to follow when factoring any polynomial.

AN IMPORTANT FORMULA

It is helpful to memorize the formula for factoring the difference of two squares. This will help you recognize when it is needed.

$$\text{Difference of two squares:} \quad x^2 - y^2 = (x + y)(x - y)$$

The difference of two squares will factor into two binomials of the form $(x + y)(x - y)$, where x is the square root of the first term and y is the square root of the second term. There is no similar rule for the sum of squares.

EXAMPLE 1

Factor: $a^2 - 25$.

Solution: The square root of a^2 is a.
The square root of 25 is 5.
Therefore:

$$a^2 - 25 = (a + 5)(a - 5)$$

The difference of two squares can be factored to the sum of their square roots times the difference of their square roots.

147

EXAMPLE 2 $x^2 - 1 = (x + 1)(x - 1)$

EXAMPLE 3 $9x^2 - 4y^2 = (3x + 2y)(3x - 2y)$

EXAMPLE 4 $16a^2 - 49 = (4a + 7)(4a - 7)$

EXAMPLE 5 $x^2y^2 - 100 = (xy + 10)(xy - 10)$

EXAMPLE 6 $y^2 + 25$ Prime—not factorable. This is the *sum* of two squares.

EXAMPLE 7 $x^3 - 4$ Prime—x^3 is *not* a square.

Now you try some.

Problem 1 $x^2 - 9$

Problem 2 $25x^2 - 1$

Problem 3 $a^2 - 121b^2$

Problem 4 $x^2 + 1$

More often than not, to factor a polynomial completely requires more than one step. The following is a suggested procedure for factoring.

SUGGESTED PROCEDURE FOR FACTORING

1.   First factor out any common monomial factor.

2.   If there are two terms, check for the difference of two squares and factor accordingly.

3.   If there are three terms, try factoring the trinomial as the product of two binomials.

EXAMPLE 8

Factor completely: $3x^4 - 27x^2$.

Solution:

common monomial factor

difference of two squares; factor again

$$3x^4 - 27x^2 = 3x^2(x^2 - 9)$$
$$= 3x^2(x + 3)(x - 3)$$

EXAMPLE 9

Factor completely: $3x^4y - 6x^3y + 3x^2y$.

Solution:

common monomial factor

three terms;
factor again

$$3x^4y - 6x^3y + 3x^2y = 3x^2y(x^2 - 2x + 1)$$
$$= 3x^2y(x - 1)(x - 1)$$

EXAMPLE 10

Factor completely: $x^4 - 16$.

Solution: This is the difference of two squares.

The square root of x^4 is x^2.

The square root of 16 is 4.

Therefore:

difference of two squares;
factor again

$$x^4 - 16 = (x^2 + 4)(x^2 - 4)$$
$$= (x^2 + 4)(x + 2)(x - 2)$$

EXAMPLE 11

Factor completely: $2x^3 + 8x^2 + 6x$.

Solution:

common monomial factor

three terms;
factor again

$$2x^3 + 8x^2 + 6x = 2x(x^2 + 4x + 3)$$
$$= 2x(x + 1)(x + 3)$$

EXAMPLE 12

Factor: $(x+3)^2 - 16$.

Solution: This is the difference of two squares.

The square root of $(x+3)^2$ is $x+3$.

The square root of 16 is 4.

Therefore:

$$(x+3)^2 - 16 = [(x+3)+4][(x+3)-4]$$
$$= (x+7)(x-1)$$

Here are a few problems for you.

Problem 5 $2x^3 - 50x$

Problem 6 $3ax^2 - 12ax - 15a$

Problem 7 $z^4 - 81$

You should now have a good understanding of factoring and be able to factor many expressions. Remember our suggested three-step procedure which, simply stated, is as follows:

1.   Factor out any common monomial factor.

2.   If there are two terms, check for difference of two squares and factor accordingly.

3.   If there are three terms, try to factor into two binomials. Before beginning the next unit, try factoring the following polynomials.

EXERCISES

Factor completely:

1. $x^2 - 64$

2. $w^2 - 81$

3. $x^2 - y^2$

4. $16x^2 - 9$

5. $3b^2 - 75$

6. $x^2 - 6x + 9$

7. $2x^2 - 2$

8. $3abc^2 - 3abd^2$

9. $x^2 + 2x - 8$

10. $x^2 - x + 7$

11. $x^3 - 36x$

12. $x^2 + 4$

13. $x^2 + 13x + 30$

14. $3r^3 - 6r^2 - 45r$

15. $x^2 + 5x - 14$

16. $2a^2b^2c^2 - 4ab^2c^2 + 2b^2c^2$

17. $5x^2y - 15xy - 10y$

18. $5x^4 + 10x^3 - 15x^2$

19. $3x^2 - 12$

20. $a^2b^2 - a^2c^2$

21. $2xy^2 - 54xy + 100x$

22. $10ab^2 - 140ab + 330a$

23. $w^2x^2y^2 + 7w^2x^2y - 18w^2x^2$

24. $2ax^2 - 2ax - 40a$

25. $4a^2 - 9b^2$

26. $2x^2 - 10x - 12$

27. $4r^3s^2 - 48r^2s^2 + 108rs^2$

28. $2y^2z + 38yz + 96z$

29. $3a^2b^5 - 3a^2b$

30. $a^2x^4 - 81a^2$

Solutions to Problems

1. $x^2 - 9 = (x + 3)(x - 3)$

2. $25x^2 - 1 = (5x + 1)(5x - 1)$

3. $a^2 - 121b^2 = (a + 11b)(a - 11b)$

4. $x^2 + 1$, prime, this is a sum of squares

5. $2x^3 - 50x = 2x(x^2 - 25) = 2x(x + 5)(x - 5)$

6. $3ax^2 - 12ax - 15a = 3a(x^2 - 4x - 5) = 3a(x - 5)(x + 1)$

7. $z^4 - 81 = (z^2 + 9)(z^2 - 9) = (z^2 + 9)(z + 3)(z - 3)$

UNIT 20

Factoring (Continued)

This unit will continue our discussion of factoring. Specifically, we will learn a technique called **factoring by grouping**. Additionally, we will learn to factor trinomials in which the coefficient of x^2 is not 1. The unit concludes with a general strategy for factoring.

FACTORING BY GROUPING

Thus far we have looked at procedures for factoring polynomials with two or three terms. We will now examine a technique for factoring a polynomial with four terms. It is called **factoring by grouping**. As the name suggests, we try grouping the four terms into pairs that have some common factor. Notice we use the word *try*. Not all polynomials with four terms can be factored using this technique. Several examples should be sufficient.

EXAMPLE 1

Factor: $ab + ac + bd + cd$.

Solution: Terms one and two have a in common.

Terms three and four have d in common.
Therefore:

$$ab + ac + bd + cd = a(b + c) + d(b + c)$$

Since we can treat $(b + c)$ as a single quantity and $(b + c)$ is common to both terms, factor it out as a common factor:

$$= (b + c)(a + d)$$

Neat technique, isn't it?

EXAMPLE 2

Factor: $x^3 + 5x^2 + 2x + 10$

Solution: $x^3 + 5x^2 + 2x + 10 = x^2(x + 5) + 2(x + 5)$ Since $(x + 5)$ is common to both terms, factor it out in front.

$$= (x + 5)(x^2 + 2)$$

EXAMPLE 3

Factor: $x^3 - 4x^2 - 3x + 12$

Solution: $x^3 - 4x^2 - 3x + 12 = x^2(x - 4) - 3(x - 4)$ Note sign change.   Since $(x - 4)$ is common; factor it out in front.

$$= (x - 4)(x^2 - 3)$$

EXAMPLE 4

Factor: $12x^2 - 9x + 8x - 6$.

Solution: $12x^2 - 9x + 8x - 6 = 3x(4x - 3) + 2(4x - 3)$ $(4x - 3)$ is common; factor it out in front.

$$= (4x - 3)(3x + 2)$$

It's your turn to try a problem.

Problem 1

Factor: $2x^3 + 4x^2 - 3x - 6$.

Solution:

FACTORING TRINOMIALS IN WHICH THE COEFFICIENT OF x^2 IS NOT 1

When the coefficient of x^2 is not 1, the problem of factoring trinomials can become far more complicated. Two procedures are presented here, both of which are similar to that of Unit 18. The first one involves some trial and error, whereas the second uses factoring by grouping.

EXAMPLE 5

Factor: $3x^2 + 4x + 1$.

Solution: Since there are only plus signs, the two binomials will both have plus signs.

$$3x^2 + 4x + 1 = (\ + \ )(\ + \ )$$

Recall from the FOIL method diagram that:
   the product of the first terms must be $3x^2$,
   the product of the last terms must be 1, and
   the sum of inner product and outer product must be $4x$.

To get $3x^2$ the factors are $3x$ and x.
To get 1 the factors are 1 and 1.
That leaves us with the possible binomial factors being $(3x + 1)(x + 1)$.

Checking to see whether the middle term is correct:
$$\begin{array}{r} x \\ +3x \\ \hline 4x \end{array}$$

Hence:

$$3x^2 + 4x + 1 = (3x + 1)(x + 1)$$

Remember that we can always verify our answer by multiplying; we should obtain the original polynomial.

EXAMPLE 6

Factor: $2x^2 - 5x + 3$.

Solution: Since the middle term only is negative, the binomial factors will both have negative signs.

$$2x^2 - 5x + 3 = (\quad - \quad)(\quad - \quad)$$

To get $2x^2$ the factors are $2x$ and x.
To get 3 the factors are 1 and 3.
That leaves us with two choices.
The possible binomial factors are either

$$(2x - 1)(x - 3) \qquad \text{or} \qquad (2x - 3)(x - 1)$$

$$\begin{array}{c} -x \\ -6x \\ \hline -7x \end{array} \qquad\qquad \begin{array}{c} -3x \\ -2x \\ \hline -5x \end{array}$$

We try each one to see which has the correct middle term. Therefore:

$$2x^2 - 5x + 3 = (2x - 3)(x - 1)$$

EXAMPLE 7

Factor: $6x^2 - x - 15$.

Solution: Since the last term is negative, the binomial factors will have different signs.

$$6x^2 - x - 15 = (\quad + \quad)(\quad - \quad)$$

To get $6x^2$ the possible factors are:

$$(6x \quad)(x \quad) \qquad \text{or} \qquad (2x \quad)(3x \quad)$$

Possible factors for 15 are $15 \cdot 1$ and $5 \cdot 3$. So the possibilities are:

$$\begin{array}{lll}
(6x + 1)(x - 15) & (6x - 1)(x + 15) & (3x + 3)(2x - 5) \\
(6x + 15)(x - 1) & (6x - 15)(x + 1) & (3x - 3)(2x + 5) \\
(6x + 3)(x - 5) & (6x - 3)(x + 5) & (3x + 5)(2x - 3) \\
(6x + 5)(x - 3) & (6x - 5)(x + 3) & (3x - 5)(2x + 3) \\
(3x + 1)(2x - 15) & (3x - 1)(2x + 15) & \\
(3x + 15)(2x - 1) & (3x - 15)(2x + 1) &
\end{array}$$

Fortunately, eight of the possibilities can be eliminated simply by observing that, since there is no common factor in the original trinomial, there can be none in any of its factored forms. For example, $(6x + 15)$ is impossible as a factor since $(6x + 15) = 3(2x + 5)$, but the original trinomial does not have a common factor of 3.

With this observation the list of binomial factors to be considered can be shortened to:

$$(6x + 1)(x - 15) \qquad (6x - 1)(x + 15) \qquad (3x + 5)(2x - 3)$$
$$(6x + 5)(x - 3) \qquad (6x - 5)(x + 3) \qquad (3x - 5)(2x + 3)$$
$$(3x + 1)(2x - 15) \qquad (3x - 1)(2x + 15)$$

Upon inspection we find that the binomial factors are:

$$(3x - 5)(2x + 3)$$

$$- 10x$$
$$9x$$

$$- x$$

Therefore:

$$6x^2 - x - 15 = (3x - 5)(2x + 3)$$

This required quite a bit of writing and is not to our liking. We prefer a different procedure that does not depend on trial and error. We think you'll find it easier to use.

In Unit 18 we considered polynomials of degree 2 where the coefficient of x^2 was 1. Now we will expand to include polynomials, $ax^2 + bx + c$, $a \neq 1$. As before, let a, b, and c be integers.

The method we will use is called "split the middle." We split the middle term, bx, into two terms, $mx + nx$, and then factor by grouping. The steps are as follows:

To factor $ax^2 + bx + c$:

Step 1: Find the product ac.

Step 2: Find two integers, m and n, such that
A. their product is ac: $mn = ac$
B. their sum is b: $m + n = b$

Step 3: Rewrite the trinomial as $ax^2 + mx + nx + c$. This is where the name "split the middle" comes from. You have split the middle term, bx, into the sum of two terms, $mx + nx$.

Step 4: Use factoring by grouping to complete the problem.

Exactly as with factoring when $a = 1$, finding the integers m and n is done by trial and error. First look for pairs of factors of ac, then find the pair that adds up to b. The same sign rules apply:

1. If the product ac is positive, then m and n will both have the *same sign*. The signs will be the same as the sign of b.

2. If ac is negative, m and n will have *opposite signs*. The sign of the number with the larger absolute value will be the same as the sign of b.

Remember, not all quadratic trinomials can be factored with integers. If no integers exist that multiply to ac and add to b, then the polynomial is prime.

EXAMPLE 8

Factor: $3x^2 + 4x + 1$

Solution: In this problem, $b = 4$ and the product $ac = 3(1) = 3$.
We need two integers where 1) the product is 3 $mn = 3$
and 2) the sum is 4 $m + n = 4$

The integers are 3 and 1. Rewrite the trinomial splitting the $4x$ into $3x$ and $1x$, then factor by grouping.

$$3x^2 + 4x + 1 = 3x^2 + 3x + 1x + 1$$
$$= 3x(x + 1) + 1(x + 1)$$
$$= (x + 1)(3x + 1)$$

Note that it does not matter in which order you write the terms when you split the middle. We also could have done it in the opposite order.

$$3x^2 + 4x + 1 = 3x^2 + 1x + 3x + 1$$
$$= x(3x + 1) + 1(3x + 1)$$
$$= (3x + 1)(x + 1)$$

EXAMPLE 9

Factor: $2x^2 - 5x + 3$

Solution: We have $b = -5$ and the product $ac = 6$.
We need two integers where 1) the product is 6 $mn = 6$
and 2) the sum is -5 $m + n = -5$

The desired integers are -2 and -3. Use them to split the middle, then factor by grouping.

$$2x^2 - 5x + 3 = 2x^2 - 2x - 3x + 3$$
$$= 2x(x - 1) - 3(x - 1) \quad \text{Be careful with the signs.}$$
$$= (x - 1)(2x - 3)$$

EXAMPLE 10

Factor: $6x^2 - x - 15$

Solution: We have $b = -1$ and $ac = -90$.

We need two integers where 1) the product is -90 $mn = -90$
and 2) the sum is -1 $m + n = -1$

The integers are 9 and -10.

$$6x^2 - x - 15 = 6x^2 + 9x - 10x - 15$$
$$= 3x(2x + 3) - 5(2x + 3) \quad \text{Watch the signs.}$$
$$= (2x + 3)(3x - 5)$$

Go back and look at Example 7 and compare the amount of work for the two methods. We think you'll see why we prefer the "split the middle" method.

It's time for you to try a couple on your own.

Problem 2   Factor: $3x^2 + 13x + 4$

Problem 3   Factor: $2x^2 + 7x - 15$

We will conclude this unit with a general strategy for factoring polynomials. It is the same strategy that was given for factoring completely at the end of Unit 18, but here we have elaborated on the second step.

Factoring Completely: A General Strategy

1.   First, factor out common monomial factors, if any.

2.   Then, factor the remaining polynomial, if possible.

   a.   For a binomial, check if it is the difference of two squares, $a^2 - b^2$. If so, factor it to $(a + b)(a - b)$. If not, the binomial is prime and you are done.

   b.   For a trinomial with $a = 1$, $x^2 + bx + c$, try to find two integers m and n where $mn = c$ and $m + n = b$. If such numbers exist, then factor into $(x + m)(x + n)$. Otherwise, the trinomial is prime and you are done.

   c.   For a trinomial with $a \neq 1$, try to find two integers, m and n, where $mn = ac$ and $m + n = b$. If these numbers exist, use them to split the middle and then factor by grouping. Otherwise, the trinomial is prime and you are done.

   d.   For a polynomial with four terms, try factoring by grouping.

EXAMPLE 11

Factor $3x^3 - 75x$

Solution: First factor out the common monomial, $3x$. The remaining polynomial is a difference of perfect squares.

$3x^3 - 75x$

Step 1: $3x(x^2 - 25)$

Step 2: $3x(x + 5)(x - 5)$

EXAMPLE 12

Factor: $4ax^2 - 2ax - 20a$

Solution: First factor out the common monomial, $2a$. The polynomial that remains is a trinomial with $a \neq 1$. Split the middle then factor by grouping.

$4ax^2 - 2ax - 20a$

Step 1: $2a(2x^2 - x - 10)$

Step 2: $2a[2x^2 - 5x + 4x - 10]$
$2a[x(2x - 5) + 2(2x - 5)]$
$2a(2x - 5)(x + 2)$

Problem 4   Factor: $5x^2 - 5$

Problem 5   Factor: $8x^2 + 8x - 30$

The following exercise list contains a variety of expressions to be factored. I suggest using the basic strategy as outlined earlier in this unit. Keep in mind that factoring the expressions completely might require repeated factoring.

EXERCISE

Factor completely:

1. $7x^2 + 10x + 3$

2. $2y^2 + 5y - 3$

3. $6x^2 + 11x + 4$

4. $3x^3 - 5x^2 - 9x + 15$

5. $4x^2 - 28x + 48$

6. $6x^2 + 13x + 6$

7. $2x^2 + 2x - 24$

8. $4x^3 - 10x^2 - 6x + 9$

9. $5x^2 - 4x - 1$

10. $4x^2 + 8x + 4$

11. $7x^2 + 13x - 2$

12. $2x^2 - 7x + 6$

13. $2y^2 - 17y + 35$

14. $7x^2 + 32x - 15$

15. $27x^2z - 3z$

16. $6z^2 + 2z - 4$

17. $x^2z - 16xz + 64z$

18. $6xw^2 + 16wx - 6x$

19. $8y + 4x + 2xy + x^2$

20. $2x^2 + 5x - 2$

21. $8x^2 + 30x - 27$

22. $xy^3 + 2y^2 - xy - 2$

23. $12x^2 - 4x - 5$

24. $x^4 - y^4$

25. $1 - a^4$

Solutions to Problems

1. $2x^3 + 4x^2 - 3x - 6 = 2x^2(x + 2) - 3(x + 2)$
$$= (x + 2)(2x^2 - 3)$$

2. $3x^2 + 13x + 4 = 3x^2 + x + 12x + 4$
$$= x(3x + 1) + 4(3x + 1)$$
$$= (3x + 1)(x + 4)$$

3. $2x^2 + 7x - 15 = 2x^2 + 10x - 3x - 15$
$$= 2x(x + 5) - 3(x + 5)$$
$$= (x + 5)(2x - 3)$$

4. $5x^3 - 5 = 5(x^2 - 1)$
$$= 5(x + 1)(x - 1)$$

5. $8x^2 + 8x - 30 = 2(4x^2 + 4x - 15)$
$$= 2[4x^2 + 10x - 6x - 15]$$
$$= 2[2x(2x + 5) - 3(2x + 5)]$$
$$= 2(2x + 5)(2x - 3)$$

UNIT 21

Solving Quadratic Equations

The objective of this unit is to illustrate the approach used to solve equations by factoring. Although the emphasis will be on solving quadratic or second-degree equations, the unit concludes with solving third-degree and higher-degree equations as well.

QUADRATIC EQUATIONS

Definition: An equation that can be written in the form $ax^2 + bx + c = 0$, with a, b, and c all real numbers, $a \neq 0$, is called a **second-degree equation** or **quadratic equation**.

Definition: The equation $ax^2 + bx + c = 0$ is called the **standard form** of the quadratic equation.

In other words, a second-degree or quadratic equation must contain a squared term, x^2, and no term with a greater power of x.

Examples of second-degree or quadratic equations are:

$$5x^2 + 1 = 0$$
$$x^2 = 25$$
$$x^2 - 5x + 1 = 7$$
$$2x + x^2 = 15x$$

whereas

$$x^3 + 2x = 5 \text{ is a third-degree equation}$$
$$x^4 = 0 \text{ is a fourth-degree equation}$$
$$x^5 + 3x^2 - 5x + 27 = 7x^3 \text{ is a fifth-degree equation}$$

161

To solve an equation is to find the values of x that satisfy the equation. With a second-degree equation there will be **at most two real solutions**, "at most" meaning there could be two unequal real solutions, one solution from two equal real solutions resulting in one value, or no real solution.

There are three ways to solve quadratic equations algebraically.

1. Taking square roots—the easiest method when it works but can only be used for "pure" quadratic equations (ones with no x term).

2. Factoring—often the fastest method, but not always possible.

3. Quadratic formula—will always work, but is long!

SOLVING QUADRATIC EQUATIONS BY TAKING SQUARE ROOTS

Definition: A quadratic equation with $b = 0$, $ax^2 + c = 0$, is called a **pure quadratic equation**.

In other words, a pure quadratic equation is one with an x^2 term and a constant term but no x term. Such an equation can always be solved as follows:

Step 1: Isolate (solve for) x^2. In other words, rewrite the equation in the form $x^2 = k$.

Step 2: Take square roots of both sides. There are three possibilities:

  a. If $k > 0$, then there are two real solutions: $x = \sqrt{k}$ and $x = -\sqrt{k}$, often written as $x = \pm\sqrt{k}$.
  b. If $k = 0$, then the only solution is $x = 0$.
  c. If $k < 0$, then the equation has no real solution.

EXAMPLE 1

Solve: $3x^2 - 48 = 0$

Solution: $\quad 3x^2 - 48 = 0$
Step 1: $\quad 3x^2 = 48$
$\quad x^2 = 16$
Step 2(a): $\quad x = \pm\sqrt{16} = \pm4.$
Answer: $\quad x = 4 \text{ or } x = -4.$

EXAMPLE 2

Solve: $x^2 + 4 = 0$

Solution: $x^2 + 4 = 0$

      Step 1: $x^2 = -4$

      Step 2(c):   Since $-4 < 0$, this equation has no real solution.

      Solution: No real answer.

EXAMPLE 3

Solve: $\dfrac{x+5}{3} = \dfrac{x}{x-2}$

Solution: This at first does not look like a quadratic equation. This is a proportion and we can cross–multiply.

$$\frac{x+5}{3} = \frac{x}{x-2}$$

Cross–multiply: $(x+5)(x-2) = 3x$

Distribute: $x^2 - 2x + 5x - 10 = 3x$

Simplify: $x^2 - 10 = 0$ Now we see it is a pure quadratic equation.

      Step 1: $x^2 = 10$

      Step 2(a): $x = \pm\sqrt{10}$

    Solution: $x = \sqrt{10}$ or $x = -\sqrt{10}$ or approximately $x = 3.162$ or $x = -3.162$.

Try a couple yourself.

Problem 1   Solve: $\dfrac{1}{3}x^2 - 4 = 8$

Problem 2   Solve: $5x(x+4) = 2(10x+15)$

Notice that in the standard form, **all terms are on the left side** of the equal sign, with only 0 on the right side.

The technique we will use to solve a **quadratic equation** by factoring involves four steps. It is based on the principle of zero products, which states that if the product of two numbers is 0, at least one of the numbers must be 0. The four steps are:

1. Write the quadratic equation in standard form. In other words, put **all** terms on the left side, with **only** 0 remaining on the **right side** of the equal sign.

2. Factor the left side of the equation.

3. Set **each** factor equal to 0.

4. Solve the new equations from step 3.

EXAMPLE 4

Solve: $x^2 - 3x = -2$.

Solution: $x^2 - 3x = -2$

Step 1. $x^2 - 3x + 2 = 0$

Step 2. $(x - 1)(x - 2) = 0$

Step 3. $x - 1 = 0$ or $x - 2 = 0$

Step 4. $x = 1$ $x = 2$

Comment: Notice that we have **two** solutions. Both must satisfy the original equation. To verify, check by substitution.

$$x^2 - 3x = -2$$

If $x = 1$: $(1)^2 - 3(1) \overset{?}{=} -2$

$1 - 3 \overset{?}{=} -2$

$-2 = -2$

If $x = 2$: $(2)^2 - 3(2) \overset{?}{=} -2$

$4 - 6 \overset{?}{=} -2$

$-2 = -2$

EXAMPLE 5

Solve: $x^2 = -6 - 5x$.

Solution: $x^2 = -6 - 5x$

Step 1. $x^2 + 5x + 6 = 0$

Step 2. $(x + 2)(x + 3) = 0$

Step 3. $x + 2 = 0$ or $x + 3 = 0$

Step 4. $x = -2$ $x = -3$

Here are two problems for you.

Problem 3

Solve: $x^2 - 2x - 48 = 0$.

Solution:

- -

Problem 4

Solve: $x^2 = 5x - 4$.

- -

Solution:

- -

Now let's try several more examples, using the same technique with slight variations.

EXAMPLE 6

Solve: $-x^2 + 2x + 3 = 0$.

- -

Solution: $-x^2 + 2x + 3 = 0$
$$x^2 - 2x - 3 = 0 \qquad \text{Note: Multiplying the entire equation by } -1 \text{ makes the factoring easier because the } x^2 \text{ term is positive.}$$
$$(x - 3)(x + 1) = 0$$
$$x - 3 = 0 \quad \text{or} \quad x + 1 = 0$$
$$x = 3 \qquad\qquad x = -1$$

- -

EXAMPLE 7

Solve: $x^2 - 4 = 6x - 13$.

- -

Solution: $x^2 - 4 = 6x - 13$
Step 1. $x^2 - 6x + 9 = 0$
Step 2. $(x - 3)(x - 3) = 0$
Step 3. $x - 3 = 0 \quad$ or $\quad x - 3 = 0$
Step 4. $x = 3 \qquad\qquad x = 3$

The two factors in step 2 are the same, resulting in two equal solutions.

- -

EXAMPLE 8

Solve: $x^2 - 25 = 0$.

Solution: $x^2 - 25 = 0$
                 $(x - 5)(x + 5) = 0$
                 $x - 5 = 0$ or $x + 5 = 0$
                      $x = 5$ $x = -5$

Alternative solution:

Since there is no x term, take the square root of
both sides. Be sure to write both answers,
plus and minus:

$x^2 - 25 = 0$
$x^2 = 25$
$x = \pm 5$

$x = 5$ or $x = -5$.

For some reason many people have difficulty in factoring the next example, where there
is no constant term. In reality it is the easiest to factor if you remember to always first factor
out any common monomial factor.

EXAMPLE 9

Solve: $x^2 - 3x = 0$.

Solution: $x^2 - 3x = 0$
               $x(x - 3) = 0$
             $x = 0$ or $x - 3 = 0$
             $x = 0$ $x = 3$

If you have difficulty factoring the next example, don't worry. In Unit 22 we will
explain the quadratic formula that allows you to solve a second-degree equation without
any factoring.

EXAMPLE 10

Solve: $5x - 3 = -2x^2$.

Solution: $5x - 3 = -2x^2$
                     $2x^2 + 5x - 3 = 0$
                  $2x^2 - x + 6x - 3 = 0$
                $x(2x - 1) + 3(2x - 1) = 0$
                    $(2x - 1)(x + 3) = 0$
                  $2x - 1 = 0$ or $x + 3 = 0$
                  $2x = 1$ $x = -3$

                  $x = \dfrac{1}{2}$

Now you try a few problems.

Problem 5	Solve:	$-x^2 + 3x + 28 = 0$
Problem 6	Solve:	$2x^2 + 10x + 20 = -30 - 10x$
Problem 7	Solve:	$2x^2 = -6x$
Problem 8	Solve:	$3x^2 - 7x - 20 = 0$

SOLVING HIGHER-DEGREE EQUATIONS BY FACTORING

Now we will consider third-degree and higher equations. Recall that with a second-degree equation there are at most two real and different solutions. Similarly, with a third-degree equation there are at most three real solutions, with a fourth-degree equation at most four real solutions, and so on.

The basic technique used to solve these equations is identical to that introduced for quadratic equations and uses the same four steps. The approach allows us to solve equations that are factorable but offers no help for those that are not factorable. A few examples should suffice.

EXAMPLE 11

Solve: $x^3 - 7x^2 + 6x = 0$.

Solution:

	Step 1.	$x^3 - 7x^2 + 6x = 0$
Factor out an x:	Step 2.	$x(x^2 - 7x + 6) = 0$
Factor again:	Step 2.	$x(x - 1)(x - 6) = 0$
	Step 3.	$x = 0$ or $x - 1 = 0$ or $x - 6 = 0$
	Step 4.	$x = 0$ $x = 1$ $x = 6$

These are the three solutions: $x = 0$ or $x = 1$ or $x = 6$

EXAMPLE 12

Solve: $2x^4 - 10x^3 - 28x^2 = 0$.

Solution: You will save yourself time and energy if you remember to factor out the greatest common monomial factor first.

$$2x^4 - 10x^3 - 28x^2 = 0$$
$$2x^2(x^2 - 5x - 14) = 0$$
$$2x^2(x - 7)(x + 2) = 0$$

$2x^2 = 0$ or $x - 7 = 0$ or $x + 2 = 0$
  $x = 0$ $x = 7$ $x = -2$

Thus, there are only three solutions to this fourth-degree equation.

Now you try a problem.

Problem 9

Solve: $x^3 - 3x^2 - 10x = 0$.

Solution:

I'll do three more examples.

EXAMPLE 13

Solve: $x^4 - 16 = 0$.

Solution:
$$x^4 - 16 = 0$$
$$(x^2 - 4)(x^2 + 4) = 0$$
$$(x - 2)(x + 2)(x^2 + 4) = 0$$

$x - 2 = 0$ or $x + 2 = 0$ or $x^2 + 4 = 0$
$\qquad x = 2 \qquad\qquad x = -2 \qquad\qquad x^2 = -4$

no real solution to this equation

Thus there are only two real solutions to this fourth-degree equation:
$x = 2$ or $x = -2$.

EXAMPLE 14

Solve: $2x(2x + 1)(x - 3)(3x - 5) = 0$.

Solution: I hope you realize that most of the work has been done for us. The entire left side of the equation has been factored with only a 0 on the right side of the equation; thus we can move immediately to Step 3:

$$2x(2x + 1)(x - 3)(3x - 5) = 0$$

$2x = 0$ or $2x + 1 = 0$ or $x - 3 = 0$ or $3x - 5 = 0$
$\quad x = 0 \qquad\qquad 2x = -1 \qquad\qquad x = 3 \qquad\qquad 3x = 5$
$$x = \frac{-1}{2} \qquad\qquad\qquad x = \frac{5}{3}$$

You try some problems.

Problem 10   Solve: $5x^4 = 5$

Problem 11   Solve: $2x^3 + 10x^2 - 18x - 90 = 0$

EXAMPLE 15

Solve: $2x^3 - 5x^2 + 7x - 23 = 0$.

Solution: This equation cannot readily be factored—grouping does not help, nor is there a common factor. Therefore the best we can do at this time is to say there are at most three real solutions (since it is a third-degree equation), but that we have no idea what they are. If you are using a graphing calculator, you can find decimal approximations for the solutions.

ALGEBRA AND THE CALCULATOR (Optional)

In Unit 3, we showed how to use the store feature of the calculator to check the solution to an equation. When an equation has multiple solutions, there is a quicker method that uses the ask feature of the table. To use this feature, you must change the table from auto (default) to ask. Press 2nd WINDOW to see the Table Setup menu. Change Indpnt from Auto to Ask by moving on top of Ask with the arrow key and then pressing Enter. Your screen should look like the one below.

Now you will enter the equation into Y= and check each solution in the table.

EXAMPLE 16

Solve and check: $2x^4 - 10x^3 - 28x^2 = 0$

Solution: In Example 12, we found the solutions to this equation are $x = 0$, $x = 7$ or $x = -2$. To use a graphing calculator to check these solutions, enter the left side of the equation into Y1. Press $\boxed{Y=}$ and enter $2x^4 - 10x^3 - 28x^2$. In the screen below, there is an exponent of 2 that is off the right margin and not visible.

Now, press $\boxed{2nd}$ $\boxed{GRAPH}$ to view the table. The table should be empty. (below left). Enter your solutions: 0, 7, –2. Each solution evaluates to 0, so each solution checks. Try a value that is not a solution, such as $x = 5$, to see what happens. Values that are not solutions do not evaluate to 0.

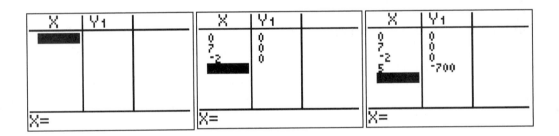

To summarize, in Unit 21 we defined, and you should now be able to identify, a second-degree equation as of the type $ax^2 + bx + c = 0$ with $a \neq 0$. Also, you should expect at most two different real solutions. One approach used to solve these equations is factoring. To find the solutions you should:

1.   Write the equation in standard form.
2.   Factor the left side.
3.   Set each factor equal to 0.
4.   Solve the new equations from step 3.

You should be able to use this same basic technique to solve third-degree and higher-degree equations. This approach allows you to solve equations that are factorable but offers no help for the many equations you are unable to factor.

Before beginning the next unit you should solve the following equations.

EXERCISES

Solve for x:

1. $x^2 + 5x - 14 = 0$

2. $x^2 + 13x + 30 = 0$

3. $6x^2 + 26 = 80$

4. $4x^2 + 8x + 4 = 0$

5. $x^2 + 5x = 0$

6. $x^2 + 2x = 8$

7. $x^3 + 5x^2 + 6x = 0$

8. $5x^2 - 5x = 0$

9. $2x^2 - 7x + 3 = 0$

10. $2x^2 + 8x + 6 = 0$

11. $z^2 + 4z - 21 = 0$

12. $10x - 10 = 19x - x^2$

13. $3x^2 + 2x = 0$

14. $2 - 2x^2 = 0$

15. $2w^2 + 7w - 4 = 0$

16. $x^3 + 3x^2 - 10x = 0$

17. $(x + 1)(x - 7)(x - 3) = 0$

18. $2x^3 - x^2 + 14x - 7 = 0$

19. $x^4 + 16x^3 + 64x^2 = 0$

20. $12x^2 + 5x - 2 = 0$

21. $\dfrac{x^2}{3} + 5 = 2$

22. $\dfrac{x}{5} = \dfrac{3x + 20}{x + 15}$

Solutions to Problems

1. $\dfrac{1}{3}x^2 - 4 = 8$

$$\dfrac{1}{3}x^2 = 12$$
$$x^2 = 36$$
$$x = \pm\sqrt{36}$$
$$x = \pm 6$$
$$x = 6 \quad \text{or} \quad x = -6$$

2. $5x(x + 4) = 2(10x + 15)$
    $5x^2 + 20x = 20x + 30$
    $\qquad 5x^2 = 30$
    $\qquad\quad x^2 = 6$
    $\qquad\quad x = \pm\sqrt{6}$
    $x = \sqrt{6} \quad \text{or} \quad x = -\sqrt{6}$

3. $x^2 - 2x - 48 = 0$
    $(x - 8)(x + 6) = 0$
    $x - 8 = 0 \qquad \text{or} \qquad x + 6 = 0$
    $\qquad x = 8 \qquad\qquad\qquad x = -6$

4. $\qquad\qquad x^2 = 5x - 4$
    $\quad x^2 - 5x + 4 = 0$
    $(x - 4)(x - 1) = 0$
    $x - 4 = 0 \qquad \text{or} \qquad x - 1 = 0$
    $\qquad x = 4 \qquad\qquad\qquad x = 1$

5. $-x^2 + 3x + 28 = 0$
    $\quad x^2 - 3x - 28 = 0$
    $(x - 7)(x + 4) = 0$
    $x - 7 = 0 \qquad \text{or} \qquad x + 4 = 0$
    $\qquad x = 7 \qquad\qquad\qquad x = -4$

6. $2x^2 + 10x + 20 = -30 - 10x$
    $2x^2 + 20x + 50 = 0$
    $2(x^2 + 10x + 25) = 0$
    $2(x + 5)(x + 5) = 0$
    $x + 5 = 0 \qquad \text{or} \qquad x + 5 = 0$
    $\qquad x = -5 \qquad\qquad\qquad x = -5$

7. $\qquad\quad 2x^2 = -6x$
    $2x^2 + 6x = 0$
    $2x(x + 3) = 0$
    $2x = 0 \qquad \text{or} \qquad x + 3 = 0$
    $\quad x = 0 \qquad\qquad\qquad x = -3$

8. $\qquad\quad 3x^2 - 7x - 20 = 0$
    $3x^2 - 12x + 5x - 20 = 0$
    $3x(x - 4) + 5(x - 4) = 0$
    $\qquad (x - 4)(3x + 5) = 0$
    $x - 4 = 0 \qquad \text{or} \qquad 3x + 5 = 0$
    $\quad x = 4 \qquad\qquad\qquad 3x = -5$
    $\qquad\qquad\qquad\qquad\qquad x = \dfrac{-5}{3}$

9. $x^2 - 3x^2 - 10x = 0$
    $x(x^2 - 3x - 10) = 0$
    $x(x - 5)(x + 2) = 0$
    $x = 0 \quad \text{or} \quad x - 5 = 0 \quad \text{or} \quad x + 2 = 0$
    $\qquad\qquad\qquad x = 5 \qquad\qquad\quad x = -2$

10.
$$5x^4 = 5$$
$$5x^4 - 5 = 0$$
$$5(x^4 - 1) = 0$$
$$5(x^2 + 1)(x^2 - 1) = 0$$
$$5(x^2 + 1)(x + 1)(x - 1) = 0$$

$x + 1 = 0$ or $x - 1 = 0$
$x = -1$ $x = 1$

11. $2x^3 + 10x^2 - 18x - 90 = 0$
$$2(x^3 + 5x^2 - 9x - 45) = 0$$
$$2[x^2(x + 5) - 9(x + 5)] = 0$$
$$2[(x + 5)(x^2 - 9)] = 0$$
$$2(x + 5)(x + 3)(x - 3) = 0$$

$x + 5 = 0$ or $x + 3 = 0$ or $x - 3 = 0$
$x = -5$ $x = -3$ $x = 3$

UNIT 22

Solving Quadratic Equations (Continued)

In this unit we will discuss how to use the quadratic formula to solve quadratic equations that cannot be factored.

Quadratic Formula

Given a quadratic equation $ax^2 + bx + c = 0$, with $a \neq 0$, the solutions are given by

$$x = \frac{-b \pm \sqrt{b^2 - 4ac}}{2a}$$

Consider the number under the radical, $b^2 - 4ac$:

1. If $b^2 - 4ac$ is negative, there are no real solutions since we cannot take the square root of a negative number.

2. If $b^2 - 4ac = 0$, there are two real and equal solutions.

3. If $b^2 - 4ac$ is positive, there are two real, distinct solutions. One is found by using the plus sign, and the other with the minus sign.

Now our technique for solving second-degree equations has been revised to:

1. Write the equation in standard form.

2. Factor the left-hand side, if possible. Otherwise, use the quadratic formula.

To practice using the quadratic formula, we will use it for this entire unit.

EXAMPLE 1

Solve: $3x^2 = x + 2$.

Solution:

$$3x^2 = x + 2$$

$$3x^2 - x - 2 = 0 \qquad \text{Rewrite in standard form.}$$

Compare with:

$$3x^2 - x - 2 = 0$$
$$a x^2 + b x + c = 0$$

a is the coefficient of the squared term,

b is the coefficient of the first-degree term,

c is the constant,

when the **equation is in standard form**.

Thus $a = 3$, $b = -1$, $c = -2$.

Now, using the formula and substituting:

$$x = \frac{-b \pm \sqrt{b^2 - 4ac}}{2a}$$

$$= \frac{-(-1) \pm \sqrt{(-1)^2 - 4(3)(-2)}}{2(3)}$$

$$= \frac{1 \pm \sqrt{1 + 24}}{6}$$

$$= \frac{1 \pm \sqrt{25}}{6}$$

$$= \frac{1 \pm 5}{6} \quad \text{since } \sqrt{25} = 5$$

Now, to find the two solutions, use first the plus and then the minus:

$$x = \frac{1 + 5}{6} \qquad \text{or} \qquad x = \frac{1 - 5}{6}$$

$$= \frac{6}{6} \qquad\qquad\qquad = \frac{-4}{6}$$

$$= 1 \qquad\qquad\qquad\quad = \frac{-2}{3}$$

Note: Had we factored $3x^2 - x - 2 = (x - 1)(3x + 2)$, we would have obtained the same solution.

EXAMPLE 2

Solve: $-5x = -3x^2 + 12$.

Solution: $-5x = -3x^2 + 12$

Compare with:

$$\underset{\textcircled{a}}{3}x^2 \underset{\textcircled{+b}}{-5} x \underset{\textcircled{+c}}{-12} = 0$$

Thus $a = 3$, $b = -5$, $c = -12$.

Substitute into the quadratic formula:

$$x = \frac{-b \pm \sqrt{b^2 - 4ac}}{2a}$$

$$= \frac{-(-5) \pm \sqrt{(-5)^2 - 4(3)(-12)}}{2(3)}$$

$$= \frac{5 \pm \sqrt{25 + 144}}{6}$$

$$= \frac{5 \pm \sqrt{169}}{6}$$

$$= \frac{5 \pm 13}{6} \quad \text{since } \sqrt{169} = 13$$

The two solutions are given by:

$$x = \frac{5 + 13}{6} \qquad \text{or} \qquad x = \frac{5 - 13}{6}$$

$$= \frac{18}{6} \qquad\qquad\qquad = \frac{-8}{6}$$

$$= 3 \qquad\qquad\qquad\qquad = \frac{-4}{3}$$

Note: Factoring:

$$3x^2 - 5x - 12 = 0$$

$$(x - 3)(3x + 4) = 0$$

would have led to the same result.

Are you ready to try one?

Problem 1

Solve: $2x^2 = 5x + 7.$

- -

- -

Now study Examples 3–6.

EXAMPLE 3

Solve: $x^2 - 2x + 4 = 0.$

- -

Solution:

$$1\,x^2 - 2\,x + 4 = 0$$

Compare: $$a\,x^2 + b\,x + c = 0$$

Thus $a = 1, b = -2, c = 4.$

Substitute into the quadratic formula:

$$x = \frac{-b \pm \sqrt{b^2 - 4ac}}{2a}$$

$$= \frac{-(-2) \pm \sqrt{(-2)^2 - 4(1)(4)}}{2(1)}$$

$$= \frac{2 \pm \sqrt{4 - 16}}{2}$$

$$= \frac{2 \pm \sqrt{-12}}{2}$$

We can stop right here and write:

No real solution, since the square root of a negative number is not real.

EXAMPLE 4

Solve: $4x^2 - 4x + 1 = 0$.

Solution:

$$\boxed{4}\, x^2 \boxed{- 4}\, x \boxed{+ 1} = 0$$

Compare:

$$\boxed{a}\, x^2 \boxed{+ b}\, x \boxed{+ c} = 0$$

Thus $a = 4$, $b = -4$, $c = 1$.

Substitute into the quadratic formula:

$$x = \frac{-b \pm \sqrt{b^2 - 4ac}}{2a}$$

$$= \frac{-(-4) \pm \sqrt{(-4)^2 - 4(4)(1)}}{2(4)}$$

$$= \frac{4 \pm \sqrt{16 - 16}}{8}$$

$$= \frac{4 \pm \sqrt{0}}{8}$$

$$= \frac{4}{8}$$

$$= \frac{1}{2}$$

There are two real and *equal* solutions.

Now let's try the same problem by factoring.

$$4x^2 - 4x + 1 = 0$$

$$4x^2 - 2x - 2x + 1 = 0$$

$$2x(2x - 1) - 1(2x - 1) = 0$$

$$(2x - 1)(2x - 1) = 0$$

$$2x - 1 = 0 \qquad \text{or} \qquad 2x - 1 = 0$$

$$2x = 1 \qquad\qquad\qquad 2x = 1$$

$$x = \frac{1}{2} \qquad\qquad\qquad x = \frac{1}{2}$$

There are two real and equal solutions.

EXAMPLE 5

Solve: $\dfrac{1}{3}x^2 - 3x - 2 = 0$.

Solution: Recall from Unit 3 that, whenever we attempted to solve an equation containing fractions, the first step was to simplify. That means to multiply the entire equation by the common denominator of all fractions.

Following that advice with this example, we should first multiply the entire equation by 3 to clear the fraction.

$$3\left(\frac{1}{3}x^2 - 3x - 2\right) = 0$$

Compare: $\underbrace{x^2}\ \underbrace{-9}x\ \underbrace{-6} = 0$
$\underbrace{a}x^2\ \underbrace{+b}x\ \underbrace{+c} = 0$

Therefore $a = 1$, $b = -9$, $c = -6$.

Substitute into the quadratic equation:

$$x = \frac{-b \pm \sqrt{b^2 - 4ac}}{2a}$$

$$= \frac{-(-9) \pm \sqrt{(-9)^2 - 4(1)(-6)}}{2(1)}$$

$$= \frac{9 \pm \sqrt{81 + 24}}{2}$$

$$= \frac{9 \pm \sqrt{105}}{2}$$

Or, if you prefer,

$$x = \frac{9 + \sqrt{105}}{2} \qquad \text{or} \qquad x = \frac{9 - \sqrt{105}}{2}$$

Note: $(9 \pm \sqrt{105})/2$ are **exact** answers to the equation. A decimal found for $\sqrt{105}$ on a calculator will only be an approximation because $\sqrt{105}$ is irrational.

EXAMPLE 6

Solve: $3x^2 + x = 6$.

Solution:

Compare: $3x^2 + x - 6 = 0$

$ax^2 + bx + c = 0$

Thus $a = 3$, $b = 1$, $c = -6$.

Substitute:

$$x = \frac{-b \pm \sqrt{b^2 - 4ac}}{2a}$$

$$= \frac{-1 \pm \sqrt{(1)^2 - 4(3)(-6)}}{2(3)}$$

$$= \frac{-1 \pm \sqrt{1 + 72}}{6}$$

$$= \frac{-1 \pm \sqrt{73}}{6}$$

Or, if you prefer,

$$x = \frac{-1 + \sqrt{73}}{6} \quad \text{or} \quad x = \frac{-1 - \sqrt{73}}{6}$$

It's your turn again.

Problem 2

Solve: $3x^2 + x - 1 = 0$.

Here are two more examples to end the unit.

EXAMPLE 7

Solve: $x^2 - 4x - 7 = 0$.

Solution: Use the quadratic formula with $a = 1$, $b = -4$, $c = -7$:

$$x = \frac{-b \pm \sqrt{b^2 - 4ac}}{2a}$$

$$= \frac{-(-4) \pm \sqrt{(-4)^2 - 4(1)(-7)}}{2(1)}$$

$$= \frac{4 \pm \sqrt{16 + 28}}{2}$$

$$= \frac{4 \pm \sqrt{44}}{2}$$

At this point you have three choices:

1. Stop and leave the answer as above.
2. Use a calculator to convert to decimal approximations.
3. Simplify the answer further as follows:

$$x = \frac{4 \pm \sqrt{4 \cdot 11}}{2}$$

$$= \frac{4 \pm 2\sqrt{11}}{2} \qquad \text{since } \sqrt{4} = 2$$

$$= \frac{4}{2} + \frac{2\sqrt{11}}{2} \qquad \text{Rewrite as seperate fractions with common denominators.}$$

$$= \frac{{}^2\cancel{4}}{\cancel{2}} + \frac{\cancel{2}\sqrt{11}}{\cancel{2}} \qquad \text{cancelling}$$

$$= 2 \pm \sqrt{11} \quad \text{is the answer in simplest form.}$$

EXAMPLE 8

Solve: $x^3 + x^2 - x = 0$.

Solution: $x^3 + x^2 - x = 0$

$x(x^2 + x - 1) = 0$

Since $x^2 + x - 1$ cannot be readily factored, use the quadratic formula.

$x = 0$ or $x^2 + x - 1 = 0$

Since $x^2 + x - 1 = 0$
then $a = 1$, $b = 1$, $c = -1$.

Substituting into the quadratic formula gives

$$x = \frac{-b \pm \sqrt{b^2 - 4ac}}{2a}$$

$$= \frac{-1 \pm \sqrt{(1)^2 - 4(1)(-1)}}{2(1)}$$

$$= \frac{-1 \pm \sqrt{5}}{2}$$

There are three solutions:

$$x = 0 \quad \text{or} \quad x = \frac{-1 + \sqrt{5}}{2} \quad \text{or} \quad x = \frac{-1 - \sqrt{5}}{2}$$

Unless they can be factored, third degree and higher degree equations are usually not solved by purely algebraic methods. There are formulas similar to the quadratic formula for solving third and fourth degree equations, but they are very complicated and seldom used. There are no general formulas to solve equations of fifth degree or higher. Equations of degree three and higher are typically solved graphically or numerically, usually with the help of technology.

ALGEBRA AND THE CALCULATOR (Optional)

Evaluating the quadratic formula on the graphing calculator can give decimal approximations for the solutions to the quadratic equation.

EXAMPLE 9

Using a calculator, find a decimal approximation for the solutions to $x^2 - 4x - 7 = 0$.

Solution: This is the quadratic equation in Example 7. Using the quadratic formula from the solution in Example 7, we have $x = \dfrac{-(-4) \pm \sqrt{(-4)^2 - 4(1)(-7)}}{2(1)}$. Without simplifying any values, this expression can be entered into a graphing calculator. The $\pm$ symbol means there are two solutions. Evaluate each solution separately. First, enter the expression with +. In the screenshot below, the expression is too wide for the display, so (–7) appears to be missing in the numerator but is included off the right margin.

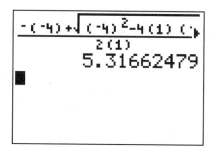

Then press 2nd ENTER to copy and paste the previous expression. Press the right arrow until the cursor is on top of the + symbol, then press the subtraction key to change it and press Enter.

Answer: $x = 5.317$ or $x = -1.317$, rounded to three decimal places

- -

You now should be able to solve any second-degree or quadratic equation. The procedure is to write the equation to be solved in standard form. Then, if the left side cannot be readily factored, compare it with $ax^2 + bx + c = 0$ to determine the values of a, b, and c. These values are then substituted into the quadratic formula

$$x = \frac{-b \pm \sqrt{b^2 - 4ac}}{2a}$$

Now try to solve the following equations using the quadratic formula.

EXERCISES

Solve for x:

1. $x^2 + 3x - 1 = 0$

2. $2x^2 - 3x - 2 = 0$

3. $4x^2 + 8x - 8 = 0$

4. $2x^2 - 3x - 1 = 0$

5. $6x^2 - 13x - 5 = 0$

6. $\frac{1}{5}x^2 - 5x + 1 = 0$

7. $10x^2 + 13x = 3$

8. $x^2 - 2x + 2 = 0$

9. $2x^2 - x = 0$

10. $9x^2 - 12x + 4 = 0$

11. $2x^2 + 7x + 9 = 0$

12. $x^2 - 2x - 10 = 0$

13. $3x^2 + x - 3 = 0$

14. $2x^2 + 3x - 4 = 0$

15. $4x^5 + 11x^4 - 3x^3 = 0$

16. $2x^4 + 2x^3 + 2x^2 = 0$

Solutions to Problems

1. $2x^2 = 5x + 7$
$2x^2 - 5x - 7 = 0$
$a = 2, b = -5, c = -7$

$$x = \frac{-(-5) \pm \sqrt{(-5)^2 - 4(2)(-7)}}{2(2)}$$

$$x = \frac{5 \pm \sqrt{25 + 56}}{4}$$

$$x = \frac{5 \pm \sqrt{81}}{4}$$

$$x = \frac{5 \pm 9}{4}$$

$$x = \frac{5 + 9}{4} \quad \text{or} \quad x = \frac{5 - 9}{4}$$

$$x = \frac{14}{7} \qquad\qquad x = \frac{-4}{4}$$

$$x = \frac{7}{2} \qquad\qquad x = -1$$

2. $3x^2 + x - 1 = 0$
$a = 3, b = 1, c = -1$

$$x = \frac{-(1) \pm \sqrt{(1)^2 - 4(3)(-1)}}{2(3)}$$

$$x = \frac{-1 \pm \sqrt{1 + 12}}{6}$$

$$x = \frac{-1 \pm \sqrt{13}}{6}$$

UNIT 23

Graphing Linear Equations in Two Variables

The purpose of this unit is to provide you with an understanding of how to graph linear equations. When you have finished the unit, you will be able to identify linear equations in two variables, distinguish them from other types of equations, and graph them.

RECTANGULAR OR CARTESIAN COORDINATE SYSTEM OF GRAPHING

We will use a rectangular or Cartesian coordinate system for all graphs. You probably remember the basic facts:

Usually the horizontal axis is labeled x.
The vertical axis is labeled y.
The origin, O, is the point where the axes cross.
Each of the four sections is called a quadrant.
Quadrants are numbered I, II, III, and IV as shown below.

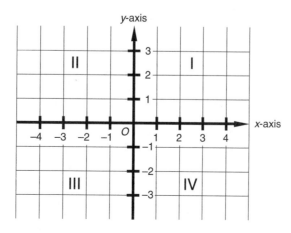

The coordinates of a point are written as an ordered pair, (x, y), where x is the number of horizontal units the point is from the origin:

to the right if x is positive,
to the left if x is negative;

and y is the number of vertical units the point is from the origin:

up if y is positive,
down if y is negative.

EXAMPLE 1

Plot: $A(2, 3)$.

Solution: 1.   Start at the origin.

2.   The first number, 2, is positive. Move two spaces to the right.

3.   Next, because the second number, 3, is positive, move three spaces up.

4.   Draw a dot and label the point.

Answer:

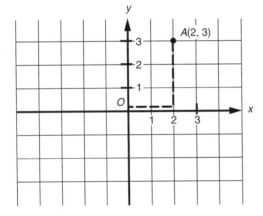

EXAMPLE 2

Plot: $B(-3, 1)$.

Solution: 1.   Start at the origin.

2.   The first number, −3, is negative. Move three spaces to the left.

3.   Next, because the second number, 1, is positive, move one space up.

4.   Draw a dot and label the point.

Answer:

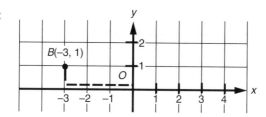

EXAMPLE 3

Plot: C(2.5, –2).

Solution: 1.   Start at the origin.

2.   Move 2.5 spaces to the right because x is 2.5.

3.   Then move two spaces down because y is –2.

4.   Draw a dot and label the point.

Answer:

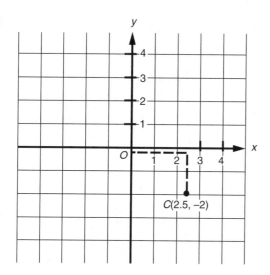

When plotting points yourself, there is no need to draw the dotted lines as I did in the examples. That was done only to show the procedure.

Now try to plot some points yourself. Use the blank grid provided. Be sure to label each point.

Problem 1 $D(-4, 3)$.

Problem 2 $E(5, 2)$.

Problem 3 $F\left(\dfrac{3}{2}, -3\right)$.

Problem 4 $G(-2, 0)$.

Problem 5 $H(0, -5)$.

Problem 6 $J(-4.5, -2)$.

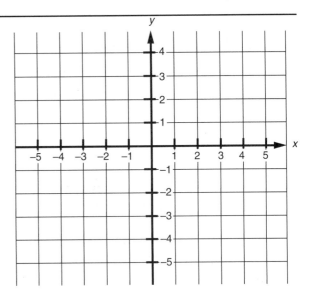

Note: Each ordered pair corresponds to one point on the graph, and for each point on the graph there is one ordered pair.

Now do a problem in which the question is reversed.

Problem 7

Find the coordinates for each of the points on the graph below:

$A(\ \ ,\ \ )$
$B(\ \ ,\ \ )$
$C(\ \ ,\ \ )$
$D(\ \ ,\ \ )$
$E(\ \ ,\ \ )$

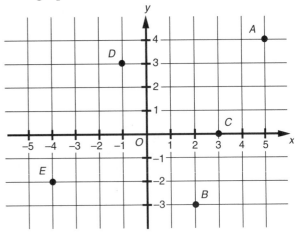

Given any ordered pair, (x, y), you should now be able to plot it. Remember that the procedure is **to start at the origin and move according to the following guide:**

First: if x is positive, move to the right;
if x is negative, move to the left.

Next: if y is positive, move up;
if y is negative, move down.

Then draw a dot and label the point.

Now we will consider what is meant by a linear equation in two variables, how we can identify such equations, and what procedure we can use to graph them.

Definition: An equation that can be written in the form $ax + by = c$, with a, b, and c, all real numbers, a and b not both zero, is **a linear equation in two variables**.

In other words, in a linear equation in two variables:

1. There are one or two variables.

2. Each variable is involved in only the four fundamental operations—addition, subtraction, multiplication, and division.

3. Neither variable is raised to any power other than 1.

4. Neither variable appears in any denominator.

5. No term contains a product of the two variables.

Here are some examples of linear equations in two variables:

$$3x + 2y = 5$$

$$\frac{1}{2}y + 7x = \frac{13}{5}$$

$$x - y = 0$$

$$\frac{x}{3} = 11y - 201$$

Here are some examples that are *not* linear equations in two variables:

$$3x^2 + 2y = 7$$

$$-5x + \sqrt{y} = 10$$

$$\frac{1}{x} - 7 = y$$

$$-13xy + y = 321$$

$$-2x - 3y + z = 15$$

Before proceeding, determine why each of the above is *not* a linear equation in two variables.

In the rest of this unit, we will consider only linear equations in two variables.

GRAPHING LINES

Now we are ready to begin graphing linear equations.

Definition: The **graph of an equation** is the set of all points whose coordinates satisfy the equation.

The graph can be thought of as a "picture" of the solution set of the equation. The graph of a linear equation is a straight line.

The basic method for graphing a linear equation is to locate and plot several points whose coordinates satisfy the equation and then connect them with a straight line. Only two points are needed to define the line, but it is a good idea to plot a third point to serve as a check. To get accurate graphs by hand, you may want even more points. There are several methods for finding points on the line.

Making a Table of Values

One basic procedure for graphing a linear equation in two variables is as follows:

1. Select a convenient value for x.

2. Substitute this value into the equation and solve for y.

3. Plot the resulting point (x, y).

4. Repeat 1–3 until you have enough points (at least two; more is better).

EXAMPLE 4

Graph: $3x - y = 5$

Solution: We recognize $3x - y = 5$ as a linear equation. Its graph will be a straight line. We will create a table to find three points on the line. We'll have a column for x, a work column for solving for y, and a column for the ordered pairs.

x	$3x - y = 5$	(x, y)

We can pick any three values for x. Because 0 is an easy number to do arithmetic with, it is often a convenient choice. Substitute it into the equation and solve for y.

x	$3x - y = 5$	(x, y)
0	$3(0) - y = 5$ $-y = 5$ $y = -5$	$(0, -5)$

Another easy number for arithmetic is 1. We'll use it as our second x-value.

x	$3x - y = 5$	(x, y)
0	$3(0) - y = 5$ $-y = 5$ $y = -5$	$(0, -5)$
1	$3(1) - y = 5$ $-y = 2$ $y = -2$	$(1, -2)$

Although two points is enough to determine a line, we'll pick a third x-value as a check; 2 seems convenient.

x	$3x - y = 5$	(x, y)
0	$3(0) - y = 5$ $-y = 5$ $y = -5$	$(0, -5)$
1	$3(1) - y = 5$ $-y = 2$ $y = -2$	$(1, -2)$
2	$3(2) - y = 5$ $-y = -1$ $y = 1$	$(2, 1)$

Now plot all three points and connect them with a straight line. Draw arrows at each end of the line to indicate that it extends indefinitely in both directions.

Answer:

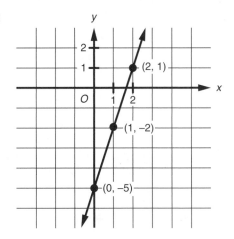

Note: In this example, we labeled the three points we plotted to help make it clear exactly how we graphed the line. Labeling the points is usually not necessary and only clutters up the graph. We will not do it in future examples.

EXAMPLE 5

Graph: $x + 2y = 6$

Solution: We recognize $x + 2y = 6$ as a linear equation, so its graph will be a straight line. We'll create a table to find three points on the line.

x	$x + 2y = 6$	(x, y)

We could pick the same three x-values we used in the last example: $x = 0, 1$, and 2. But solving for y will involve dividing by 2 in the last step. Using $x = 1$ will lead to y being a fraction. There is nothing wrong with that, but many people prefer to plot integers. It is convenient in this problem to choose only even numbers for x, say $x = 0, 2$, and 4.

x	$x + 2y = 6$	(x, y)
0	$0 + 2y = 6$ $2y = 6$ $y = 3$	$(0, 3)$
2	$2 + 2y = 6$ $2y = 4$ $y = 2$	$(2, 2)$
4	$4 + 2y = 6$ $2y = 2$ $y = 1$	$(4, 1)$

Plot the three points and connect them with a straight line.

Answer:

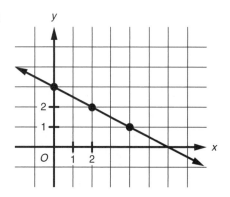

Have you noticed that in each of the last two examples, we ended up solving essentially the same equation three times? Wouldn't it be faster if we could solve the equation only once? Before selecting values for x, we will solve the equation once for y. This will be our strategy for the next few examples. If you need a refresher on how to solve an equation for y, refer to Unit 6 on solving literal equations.

Our slightly modified method for making our table of values is as follows.

1. First solve the equation for y.

2. Create a table of values

   a. Select several convenient values for x (at least two; more is better).

   b. Substitute each one into the equation and evaluate y.

3. Plot the resulting points and connect with a straight line.

EXAMPLE 6

Graph: $2x + y = 3$.

Solution: $2x + y = 3$ is a linear equation; the graph will be a straight line.

1. Solve the equation for y. $\quad 2x + y = 3$

$$y = -2x + 3$$

2. Create a table of values.

Choose convenient values for x and substitute into the solved equation to find y. We will again use 0, 1, and 2 as x-values.

x	$y = -2x + 3$	(x, y)
0	$y = -2(0) + 3$ $= 3$	$(0, 3)$
1	$y = -2(1) + 3$ $= 1$	$(1, 1)$
2	$y = -2(2) + 3$ $= -1$	$(2, -1)$

You see how having the equation solved for y makes the evaluation of the y-values easier.

3. Plot the points and connect with a straight line.

Answer:

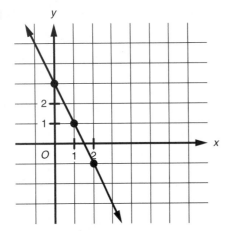

We'll do another example.

═══

EXAMPLE 7

Graph: $2x + 3y = 21$.

Solution: $2x + 3y = 21$ is a linear equation; the graph will be a straight line.

1.   Solve the equation for y.

$$2x + 3y = 21$$
$$3y = -2x + 21$$
$$y = \frac{-2}{3}x + 7$$

2.   Create a table of values.

Because the coefficient of x is $-\dfrac{2}{3}$, it will be convenient to choose x-values that

are divisible by 3. This will lead to integer y-values and make the points easier to plot. We will use 0, 3, and 6 as x-values.

x	$y = -\dfrac{2}{3}x + 7$	(x, y)
0	$y = -\dfrac{2}{3}(0) + 7$ $= 7$	$(0, 7)$
3	$y = -\dfrac{2}{3}(3) + 7$ $= 5$	$(3, 5)$
6	$y = -\dfrac{2}{3}(6) + 7$ $= 3$	$(6, 3)$

3. Plot the points and connect with a straight line.

Answer:

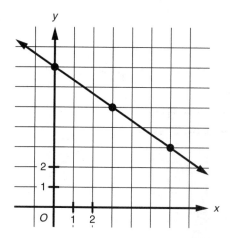

Now you graph a linear equation.

Problem 8

Graph: $5x + 2y = 10$.

Solution: $5x + 2y = 10$ is a linear equation; the graph will be a straight line.

1. Solve the equation for y.

2. Create a table of values:

x	y	(x, y)
0		(,)
2		(,)
4		(,)

Have you noticed that by first solving the equation for y, it is easier to decide what values of x are most convenient; that is, you know in advance if you want to select numbers divisible by 3 as in Example 7 or divisible by 2 as in Problem 8. Also, once you have selected the values for x, you need only to use substitution to find the corresponding y-values.

3. Plot the three points on the blank grid provided and connect them with a straight line.

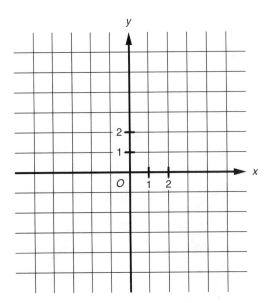

Reminder: Be sure to put the arrows at each end to indicate that the line continues on in both directions.

Intercept Method

Two points on a line deserve special attention. These are the points where the line intersects the coordinate axes. The point where the line intercepts the x-axis has coordinates $(a, 0)$. The x coordinate of that point, a, is called the **x-intercept.** The x-intercept can always be found by substituting $y = 0$ into the equation and solving for x.

Similarly, the point where the graph intersects the y-axis has coordinates $(0, b)$. The y coordinate of that point, b, is called the **y-intercept.** The y-intercept can always be found by substituting $x = 0$ into the equation and solving for y.

Since two points are all that is needed to determine a line, a line may be graphed by finding the values of the two intercepts, graphing them on the appropriate axes, and connecting them with a straight line. (If you want, you can graph a third point using some convenient x-value as a check.) This is called the intercept method of graphing a line.

EXAMPLE 8

Find the x- and y-intercepts of the line $3x - 4y = 12$ and graph the line.

Solution: The x-intercept is the value of x when $y = 0$.

By substitution: $3x - 4(0) = 12$

$$3x = 12$$
$$x = 4$$

The x-intercept is 4.

The y-intercept is the value of y when $x = 0$.

By substitution: $3(0) - 4y = 12$

$$-4y = 12$$
$$y = -3$$

The y-intercept is -3.

To graph the line, note that an x-intercept of 4 means the line passes through the x-axis at (4, 0). A y-intercept of -3 means the line passes through the y-axis at (0, -3). Plot these two points and connect them with a straight line.

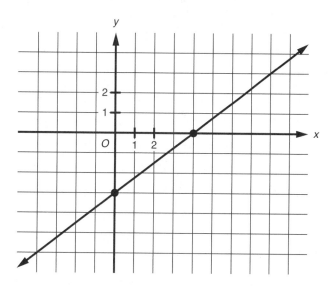

Try it yourself.

Problem 9

Find the x- and y-intercepts of the line $3x + 5y = 30$ and graph the line.

Solution:

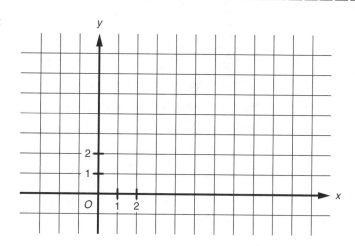

Slope-Intercept Method

There is a third way to graph lines, called the "slope-intercept method," that is often quicker and easier than making a table of values. The "intercept" refers to the y-intercept, which we've already discussed. Before we can use the method, we need to know what the "slope" of a line is.

The slope of a line is a measure of its direction and steepness. The symbol for slope is m, and it is defined as the ratio $m = \dfrac{rise}{run}$, where the "rise" and the "run" are the vertical and horizontal distances between any two points on the line, as shown in the diagram.

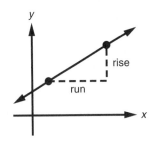

If the rise is large compared to the run, the line will have a steep slope. If the rise is small compared to the run, the line will be less steep. By definition, lines that rise—go "uphill" from left to right—have positive slopes. Lines that fall—go "downhill" from left to right—have negative slopes.

EXAMPLE 9

Find the slope of each line shown.

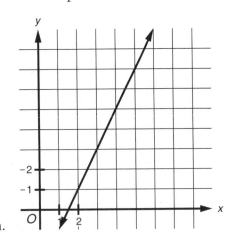

a.

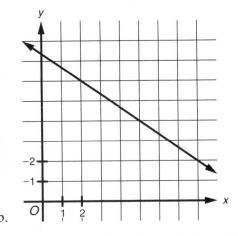

b.

Solution: To find the slope, we will pick any two points on the line and use them to find the rise and run.

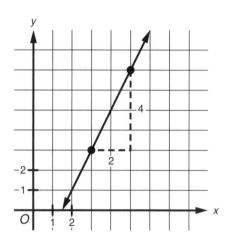

a.   Using the two points shown above, we have a rise of 4 and a run of 2. This gives a slope of $m = \dfrac{4}{2} = \dfrac{2}{1}$, or just 2. A helpful way to interpret this is that as we move along this line, every time x increases (moves right) by 1, y increases (goes up) by 2.

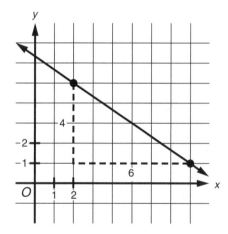

b.   Using the two points shown above, we have a rise of −4; the rise is a negative number because, moving from left to right, we are actually "falling" instead of rising. The run is 6, always positive if we move from left to right. This gives a slope of $m = \dfrac{-4}{6} = -\dfrac{2}{3}$. One way to interpret this is that as we move along this line, every time x increases (moves right) by 1, y decreases (goes down) by $\dfrac{2}{3}$. An equivalent interpretation that is more convenient for graphing is: as x increases by 3, y decreases by 2.

Any linear equation that involves y can be written in the form $y = mx + b$, where m and b are numbers, by solving the equation for y. This is called the "slope-intercept form" of the equation of a line. In this form, m is the slope of the line and b is the y-intercept.

> **Slope-Intercept Equation of a Line**
>
> $$y = mx + b$$
>
> m is the slope
>
> b is the y-intercept

EXAMPLE 10

Write the equation of the line $2x + 5y = 15$ in slope-intercept form and identify the slope and the y-intercept.

_____

Solution: We need to solve for y.

$$2x + 5y = 15$$
$$5y = -2x + 15$$
$$y = -\frac{2}{5}x + 3$$

By comparing this to $y = mx + b$, we see that the slope is $m = -\dfrac{2}{5}$ and the y-intercept is $b = 3$.

_____

We are now ready to graph lines using the slope-intercept method.

1.   Solve for y and write the equation in slope-intercept form: $y = mx + b$. Write the slope as a fraction: $m = \dfrac{c}{d}$; if the slope is negative, include the sign in the numerator.

2.   Plot the y-intercept on the y-axis: $(0, b)$.

3.   Starting at the y-intercept, count up c and right d and plot a new point. Note that if c is negative, "up c" means "down $|c|$."

4.   Repeat step 3 as necessary to find more points. You can also start at the y-intercept and work backwards: down c and left d.

5.   Connect the points with a straight line.

Some examples should help clarify this. It is easier than it sounds at first.

EXAMPLE 11

Graph the line $2x + 5y = 15$.

_____

Solution: 1.   Solve for y (this was done in Example 10): $y = \dfrac{-2}{5}x + 3$

Note this time we wrote the slope with the negative sign in the numerator.

2.   Plot the y-intercept at $(0, 3)$ on the y-axis.

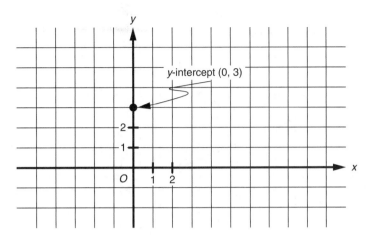

3. The slope is $\dfrac{-2}{5}$ which we can interpret as "down 2, right 5."

Starting from the y-intercept, count down 2 and right 5 to find a second point on the line.

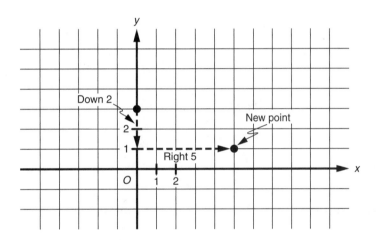

4. We can repeat step 3, down 2 and right 5, from the new point to get a third point. We can also go back to the y-intercept and work backwards, up 2 and left 5, to get a point in the second quadrant. This is optional, but more points will usually give you a better graph when graphing by hand.

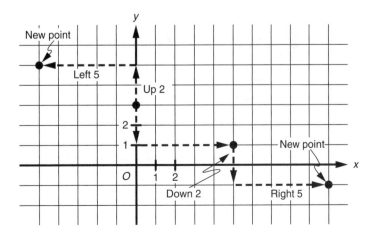

5.   Connect the points with a straight line.

Answer:

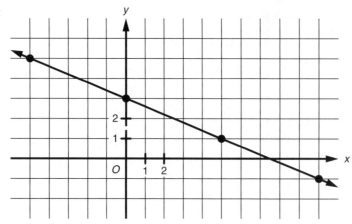

Here is a second example. This time we'll combine some steps.

EXAMPLE 12

Graph the line $2x - y = 5$.

Solution: Solve for y: $y = 2x - 5$

$$y = \frac{2}{1}x - 5$$

Note we rewrote the slope as a fraction with 1 in the denominator.

The y-intercept will be graphed at $(0, -5)$. From there, we will find new points by going up 2 and right 1, or by going down 2 and left 1.

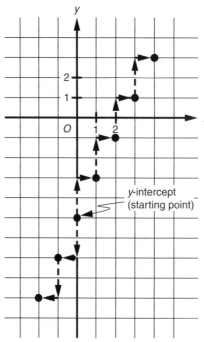

Finally, connect the points with a straight line.

Answer:

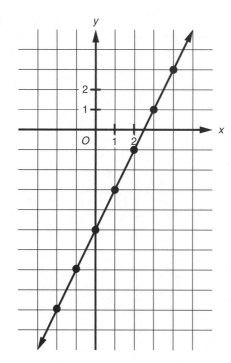

Now you try a couple. Note that the first one is already in slope-intercept form.

Problem 10

Graph the line $y = \dfrac{3}{4}x - 2$

Solution:

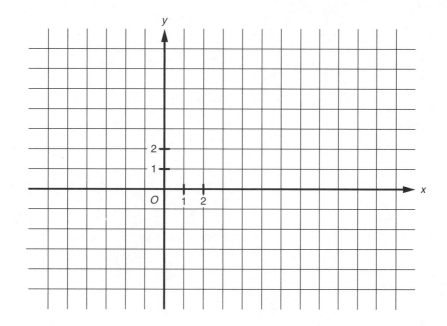

Problem 11

Graph the line $2x + 3y = 9$

Solution:

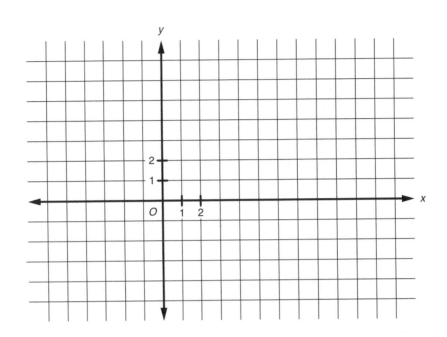

TWO SPECIAL TYPES OF LINEAR EQUATIONS

There are two special types of linear equations in which only a single variable, either x or y, is involved. The graph of such an equation will be either a vertical or a horizontal line, depending on which variable is present.

EXAMPLE 12

Graph: $x = 3$.

Solution: Recall that a linear equation can be written in the form $ax + by = c$ with a and b not both zero.

$x = 3$ can be written as $x + 0y = 3$.

Therefore $x = 3$ is classified as a linear equation; its graph will be a straight line. Locate three points whose coordinates satisfy the equation. The equation $x = 3$ says all x-values must be 3. Because y has a coefficient of 0, the y-values can be any number. So, three points whose coordinates satisfy the equation are (3, 2), (3, 5), and (3, 0).

To verify (3, 2): $x + 0y = 3$

$3 + 0(2) = 3$

$3 = 3$

Thus point (3, 2) is on the line.

Answer:

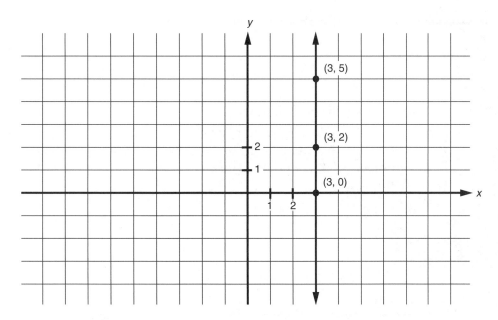

Observe: The graph of $x = 3$ is a vertical line.

Each point on the line has an x-coordinate of 3.

The x-intercept is 3.

There is no y-intercept.

The slope is undefined.

> The graph of a linear equation $x = r$, where r is a real number, is a *vertical line* with x-intercept of r.

EXAMPLE 13

Graph: $x = -4$.

Solution: The graph of a linear equation $x = -4$ is a vertical line with x-intercept of -4.

Answer:

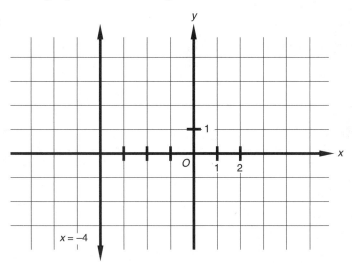

EXAMPLE 14

Graph: $y = 2$.

Solution: By reasoning similar to Example 12, $y = 2$ can be written as $0x + y = 2$.

Thus $y = 2$ is also a linear equation; its graph will be a straight line. The equation $y = 2$ says all y-values must be 2. Because x has a coefficient of 0, the x-values can be any number. So, three points whose coordinates satisfy the equation are $(5, 2)$, $(-3, 2)$, and $(0, 2)$.

Answer:

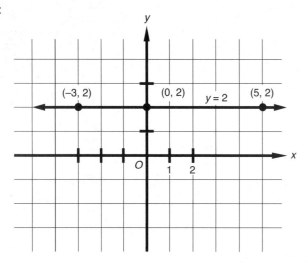

Observe: The graph of $y = 2$ is a horizontal line.

Each point on the line has a y-coordinate of 2.

The y-intercept is 2.

There is no x-intercept.

The slope is zero.

The graph of a linear equation $y = p$, where p is a real number, is a *horizontal line* with y-intercept of p.

You do a final problem.

Problem 12

Graph: $\left.\begin{array}{l} x = 5 \\ y = -3 \\ y = 2x - 1 \end{array}\right\}$ on the same set of axes.

Solution:

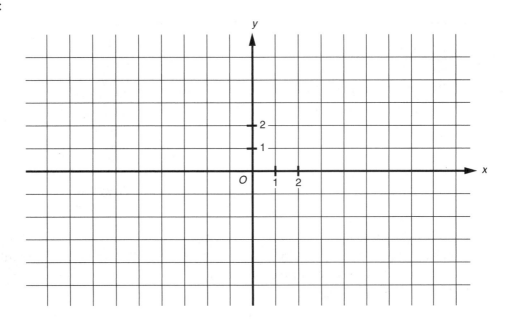

ALGEBRA AND THE CALCULATOR (Optional)

Graphing Lines

So that we start at the same place, after pressing the $\boxed{\text{ON}}$ key,

    press $\boxed{\text{2nd}}$ $\boxed{\text{MODE}}$ $\boxed{\text{CLEAR}}$ and

    press $\boxed{\text{2nd}}$ $\boxed{\text{TblSet}}$ to view the table setup menu.

Now assign   TblStart = 0   by pressing $\boxed{0}$ $\boxed{\text{ENTER}}$.

                Δ Tbl = 1   by pressing $\boxed{1}$ $\boxed{\text{ENTER}}$.

                Indpnt: Ask   by highlighting Ask and pressing $\boxed{\text{ENTER}}$.

                Depend: Auto by highlighting Auto and pressing $\boxed{\text{ENTER}}$.

You should see the following:

```
TABLE SETUP
 TblStart=0
 ΔTbl=1
Indpnt: Auto ASK
Depend: Auto Ask
```

To return to the home screen, press $\boxed{\text{2nd}}$ $\boxed{\text{QUIT}}$.

For now we will use the standard viewing window for our graphs. Its dimensions are [–10, 10] by [–10, 10]. That is, the *x*-axis will be shown from –10 to 10 and the *y*-axis will be shown from –10 to 10 as pictured at the right.

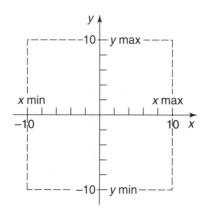

To use the standard viewing window, press Y= CLEAR ZOOM 6, and a blank standard viewing window should appear on the screen.

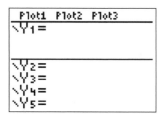

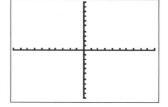

Press WINDOW to confirm that the viewing window's dimensions agree with the screen shown here. If they don't, reset to match.

Return to the home screen by pressing 2nd QUIT.

```
WINDOW
 Xmin=-10
 Xmax=10
 Xscl=1
 Ymin=-10
 Ymax=10
 Yscl=1
 Xres=1
```

Creating a Table of Values and the Graph of a Line

EXAMPLE 15

Graph: $y = 1.3x + 2.75$ using a graphing calculator.

Solution: This example is the same as what you have been doing except that the equation includes decimals. Thus, manual calculations would be tedious and time-consuming, and the graph would be an approximation. We will do it on a calculator.

To enter the equation, press the following keys.

Y= CLEAR 1 · 3 X,T,θ,n + 2 · 7 5 ENTER.

You should see the following.

```
Ploti Plot2 Plot3
\Y1◼1.3X+2.75
\Y2=
\Y3=
\Y4=
\Y5=
\Y6=
\Y7=
```

To find points on the line, press 2nd TABLE.

Select a value for x.   Let $x = 0$.

To find y, press 0 ENTER.

Repeat by entering 1, 2, and 3.

Reminder: The (−) key is used to denote a negative number. An error message will appear if the subtraction key − is used by mistake. Finish the above example by entering −1, −2, and −3.

You should see the following.

```
X      Y1
0      2.75
1      4.05
2      5.35
3      6.65
-1     1.45
-2     .15
-3     -1.15
X=
```

To see the graph, press GRAPH. Surprisingly easy, wasn't it?

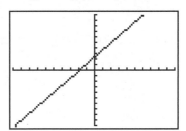

EXAMPLE 16

Graph: $y = -\frac{1}{2}x + 4$ using a graphing calculator.

Find: a. y when $x = 3.74$ using the CALC key.

b. the y-intercept using the TRACE key.

c. the x-intercept using the CALC key.

Solution: Create a table of values for the equation using the following keystrokes:

$\boxed{\text{Y=}}$ $\boxed{\text{CLEAR}}$ $\boxed{(-)}$ $\boxed{\text{ALPHA}}$ $\boxed{\text{Y=}}$ $\boxed{1}$ $\boxed{1}$ $\boxed{)}$ $\boxed{2}$ $\boxed{)}$ $\boxed{\text{X,T,θ,n}}$ $\boxed{+}$ $\boxed{4}$ $\boxed{\text{ENTER}}$.

To see various x- and y-values, press $\boxed{\text{2nd}}$ $\boxed{\text{TABLE}}$.

To see the graph, press $\boxed{\text{GRAPH}}$.

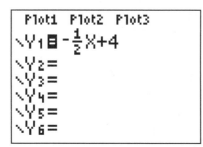 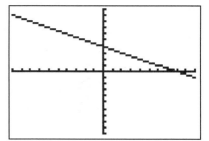

a. With the graph on the screen, to find y when $x = 3.74$ using the $\boxed{\text{CALC}}$ key, press $\boxed{\text{2nd}}$ $\boxed{\text{CALC}}$ $\boxed{1}$ $\boxed{3}$ $\boxed{\cdot}$ $\boxed{7}$ $\boxed{4}$ $\boxed{\text{ENTER}}$. The equation appears at the upper right, the x- and y-values of 3.74 and 2.13 are at the bottom, and a star appears on the graph at the point with coordinates (3.74, 2.13).

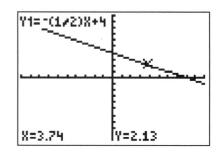

b. To find the y-intercept using the $\boxed{\text{TRACE}}$ key, press $\boxed{\text{TRACE}}$.

Press 0 Enter. Using the arrow keys to move the cursor also works but is inaccurate if you cannot move the cursor to exactly the right value. Just type in the x-value, press Enter, and the cursor will jump to the value.

The y-intercept, in this example, $y = 4$, appears at the bottom of the screen.

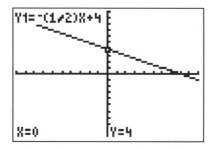

c. To find the x-intercept, also called a zero, using the $\boxed{\text{CALC}}$ key requires a bit more work. Press $\boxed{\text{2nd}}$ $\boxed{\text{CALC}}$ $\boxed{2}$.

At the <u>Left Bound?</u> prompt, using only the $\boxed{\blacktriangleright}$ and $\boxed{\blacktriangleleft}$ arrows, move the cursor to a point just left of the x-intercept and press $\boxed{\text{ENTER}}$. If you find it difficult to decide where to place the cursor, keep in mind that both the x- and y-values will be positive.

At the <u>Right Bound?</u> prompt, move the cursor to a point just to the right of the x-intercept. Here x will be positive and y will be negative; then press ENTER.

At the <u>Guess?</u> prompt, move the cursor between the boundaries shown on the screen by the two arrows, ▶ ◀, and press ENTER. The answer appears at the bottom of the screen: Zero $x = 8$ $y = 0$.

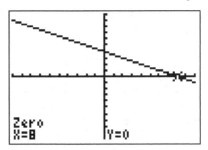

Before continuing, we suggest you redo Examples 4–11 for practice with your calculator. Also, when entering functions, experiment with the DEL and INS keys.

EXAMPLE 17

Graph $2x + 5y = 60$ using a graphing calculator.

Solution: Before the equation can be entered into the calculator, it must be solved for y as follows:

$$2x + 5y = 60$$
$$5y = -2x + 60$$
$$y = -\frac{2}{5}x + 12$$

Press Y= (−) ALPHA Y= 1 2) 5) X,T,θ,n + 1 2 ENTER. This time, however, when we press GRAPH , only a small portion of the line is visible in the standard viewing window.

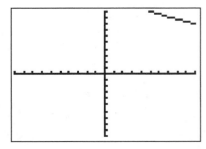

Obviously the viewing window needs to be enlarged. Several methods are available.

One method is to press ZOOM 3 ENTER.

If you are curious about the new dimensions, press WINDOW.

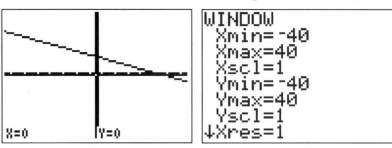

A second method is to manually reset the dimensions of the window. For example, you might find the x- and y-intercepts, which are 30 and 12, respectively, and reset the window's dimensions accordingly.

To reset the dimensions of the window, press WINDOW and reset Xmax to 35 to accommodate the x-intercept of 30, and reset Ymax to 15 to accommodate the y-intercept of 12. Press GRAPH to view the graph in the new enlarged viewing window.

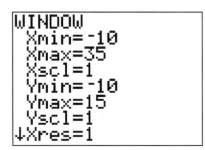 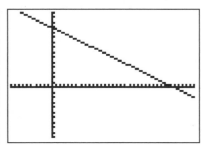

Now we encourage you to do a bit of experimenting with the settings and the various keys. For instance, press 2nd FORMAT, select various options, and observe the results on a graph. Or, change the viewing window's dimensions again. ZOOM 6 will always take you back to the standard viewing window. Or, use TRACE to move along the line and observe what happens when you reach the edge of the viewing window.

You should now be able to recognize and graph linear equations in two variables by one of three methods: creating a table of values, graphing the x- and y-intercepts, or graphing using the slope-intercept method.

To create a table of values:

> solve the equation for y;
> select three convenient values for x and solve for y by substitution.

To graph using the x- and y-intercepts:

> locate the y-intercept, which is the value of y when $x = 0$;
> locate the x-intercept, which is the value of x when $y = 0$;
> plot and connect the two points with a straight line.

To graph using the slope-intercept method:

solve the equation for y;
locate the y-intercept, which is the constant term, and plot this value on the y-axis;
locate the slope, which is the coefficient of x, and use the slope to plot points up or down and to the right of the y-intercept;
connect the points with a straight line.

Before beginning the next unit you should graph the following equations.

EXERCISES

Graph each equation. Identify the x-intercept, the y-intercept, and the slope.

1. $x + y = 8$

2. $3x - 4y = 12$

3. $7x + y = 10$

4. $y = 5$

5. $y - 2x = 0$

6. $x + 3y = 21$

7. $y = \dfrac{2}{7}x + 4$

8. $x = -3$

9. $-3x = 2y + 4$

10. $y = \dfrac{2}{5}x - 1$

C11. Graph $y = \dfrac{3}{4}x - 5$ using a graphing calculator.

C12. Using a graphing calculator and given $y = 1.23x + 5.76$,

    a. Graph the linear equation.

    b. Find the y-intercept.

    c. Find the x-intercept.

    d. Find y so that the point $(-1.12, y)$ is on the line.

- -

Solutions to Problems

1–6.

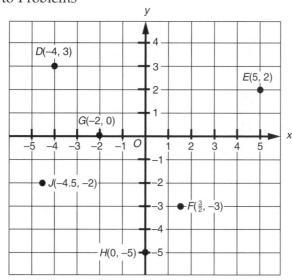

7. A(5, 4), B(2, –3), C(3, 0), D(–1, 3), E(–4, –2)

8.

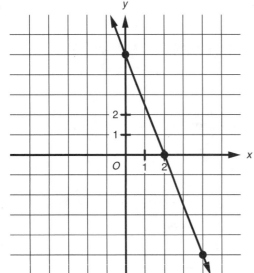

9.

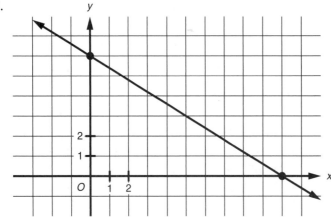

10.

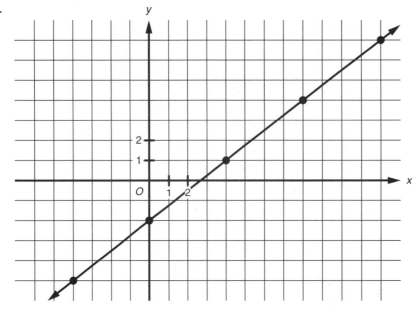

11.

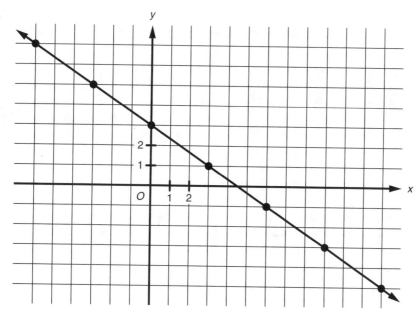

12.

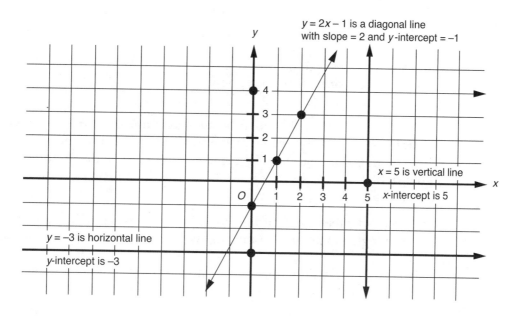

$y = 2x - 1$ is a diagonal line
with slope = 2 and y-intercept = −1

$x = 5$ is vertical line

x-intercept is 5

$y = -3$ is horizontal line

y-intercept is −3

UNIT 24

Graphing Quadratic Equations

In this unit you will learn to recognize quadratic equations in two variables and to graph them using a minimum number of well-chosen points.

> Definition: $y = ax^2 + bx + c$, with a, b, and c being real numbers, $a \neq 0$, is called a **quadratic** or **second-degree equation in two variables.** When written as shown, with y alone on the left and the powers of x in descending order on the right, the equation is said to be in **standard form**.

In other words, a second-degree equation in two variables must contain two variables, a squared term, x^2, and no higher exponent. In this unit we will consider only second-degree or quadratic equations.

The basic method for graphing a quadratic equation is to make a table of values. Recall from Unit 23 that two points are enough to graph a linear equation. Graphing a quadratic equation requires a minimum of three well-chosen points. To get an accurate graph by hand, we prefer to have five to seven points. Let's look at a representative example.

EXAMPLE 1

Graph: $y = x^2 - 2x - 3$

Solution: We will make a table of values. For this example, we'll use the integer x-values from $x = -2$ to $x = 4$. After this example, we'll discuss how to choose appropriate x-values to graph.

216

x	$y = x^2 - 2x - 3$	(x, y)
–2	$y = (-2)^2 - 2(-2) - 3$ $= 5$	(–2, 5)
–1	$y = (-1)^2 - 2(-1) - 3$ $= 0$	(–1, 0)
0	$y = (0)^2 - 2(0) - 3$ $= -3$	(0, –3)
1	$y = (1)^2 - 2(1) - 3$ $= -4$	(1, –4)
2	$y = (2)^2 - 2(2) - 3$ $= -3$	(2, –3)
3	$y = (3)^2 - 2(3) - 3$ $= 0$	(3, 0)
4	$y = (4)^2 - 2(4) - 3$ $= 5$	(4, 5)

Now plot the points and connect them with a smooth curve.

Answer:

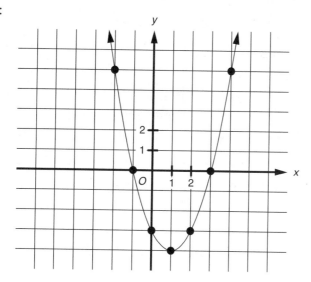

- -

Be sure to notice that every ordered pair whose coordinates satisfy the equation, $y = x^2 - 2x - 3$, must correspond to some point on the curve. And every point on the curve must have coordinates that satisfy the equation.

As you can see, the graph of a quadratic equation in two variables is not a line. Instead, it is a ⌣ –shaped curve called a **parabola**. The graphs of parabolas of the form $y = ax^2 + bx + c$ all share some properties that are helpful to know when graphing them.

Properties of Parabolas
$$y = ax^2 + bx + c$$

1. The graph is a smooth ⌣-shaped curve. It does not come to a point, V.
2. If a, the coefficient of the squared term, is positive, the curve opens up: ⌣.
    If a is negative, the curve opens down: ⌢.
3. The y-intercept is c, the constant. Remember that the y-intercept is the value of y when $x = 0$; thus $y = a(0)^2 + b(0) + c = c$.
4. There are **at most two x-intercepts**, "at most" meaning there can be two, one, or none. The x-intercepts are the values of x when $y = 0$; thus the solution to $0 = ax^2 + bx + c$ yields the x-intercepts.
5. The low point on the curve (or the high point if the curve opens down) is called the **vertex**.
6. The curve is symmetric to a vertical line through the vertex. The vertical line is called the **axis of symmetry**. Its equation is $x = \dfrac{-b}{2a}$.
7. The vertex is located at the point where $x = \dfrac{-b}{2a}$. Find the y-value of the vertex by substituting the x-value into the quadratic equation.

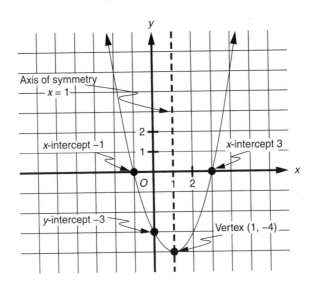

Based on the properties of parabolas shown above, we suggest the following method for graphing equations of the form $y = ax^2 + bx + c$.

GRAPHING A QUADRATIC EQUATION IN TWO VARIABLES

1. Write the quadratic equation in standard form.
2. Find the x-value of the axis of symmetry: $x = -\dfrac{b}{2a}$.
3. Make a table of values including the axis of symmetry and three integral x-values on each side of the axis, a total of seven points. (If the y-values become very large, we may settle for fewer x-values.)

4. Plot the points and connect them with a smooth curve.
5. Check that your graph matches the equation.

  - Is it $\smile$-shaped (not a line)?
  - Does it open in the right direction, up if $a > 0$, down if $a < 0$?
  - Does it pass through the y-intercept, $(0, c)$?

Steps 2 and 3 explain how we chose the x-values for Example 1. For $y = x^2 - 2x - 3$, the axis of symmetry is $x = -\dfrac{(-2)}{2(1)}$, or $x = 1$. We wanted our table of values to include $x = 1$ and three integral x-values on each side. Thus we used $x = -2, -1, 0, 1, 2, 3,$ and 4.

Let's see a second example.

EXAMPLE 2

Graph: $y = x^2 - 5x + 4$.

- -

Solution: 1. The equation is already in standard form (this is pretty common).

2. The axis of symmetry is $x = -\dfrac{b}{2a} = -\dfrac{(-5)}{2(1)}$, or $x = 2.5$.

3. Make a table of values including the axis of symmetry and three integral x-values on each side.

x	$y = x^2 - 5x + 4$	(x, y)
0	$y = (0)^2 - 5(0) + 4$ $= 4$	$(0, 4)$
1	$y = (1)^2 - 5(1) + 4$ $= 0$	$(1, 0)$
2	$y = (2)^2 - 5(2) + 4$ $= -2$	$(2, -2)$
2.5	$y = (2.5)^2 - 5(2.5) + 4$ $= -2.25$	$(2.5, -2.25)$
3	$y = (3)^2 - 5(3) + 4$ $= -2$	$(3, -2)$
4	$y = (4)^2 - 5(4) + 4$ $= 0$	$(4, 0)$
5	$y = (5)^2 - 5(5) + 4$ $= 4$	$(5, 4)$

    4. Plot the points and connect them with a smooth curve.

Answer:

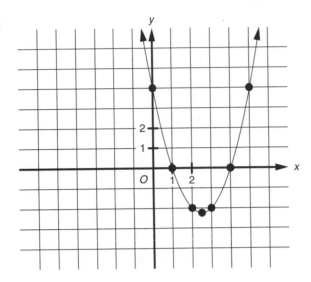

    5. Check that the graph matches the equation.

        Is it ∪-shaped? ✓

        Does it open in the right direction? $a = 1 > 0$, should open up. ✓

        Does it pass through the y-intercept, $(0, 4)$? ✓

EXAMPLE 3

Graph: $y = -2x^2 - 8x - 9$

Solution: 1. The equation is already in standard form.

        2. The axis of symmetry is $x = -\dfrac{b}{2a} = -\dfrac{(-8)}{2(-2)}$, or $x = -2$.

        3. Make a table of values including the axis of symmetry and three integral x-values on each side.

x	$y = -2x^2 - 8x - 9$	(x, y)
-5	$y = -2(-5)^2 - 8(-5) - 9$ $= -19$	$(-5, -19)$
-4	$y = -2(-4)^2 - 8(-4) - 9$ $= -9$	$(-4, -9)$
-3	$y = -2(-3)^2 - 8(-3) - 9$ $= -3$	$(-3, -3)$
-2	$y = -2(-2)^2 - 8(-2) - 9$ $= -1$	$(-2, -1)$

-1	$\begin{aligned} y &= -2(-1)^2 - 8(-1) - 9 \\ &= -3 \end{aligned}$	$(-1, -3)$
0	$\begin{aligned} y &= -2(0)^2 - 8(0) - 9 \\ &= -9 \end{aligned}$	$(0, -9)$
1	$\begin{aligned} y &= -2(1)^2 - 8(1) - 9 \\ &= -19 \end{aligned}$	$(1, -19)$

Note that two of our y-values are -19. These would be inconvenient to graph on a graph with a scale of 1, so we will omit them and settle for just five points.

4.   Plot the points and connect them with a smooth curve.

Answer:

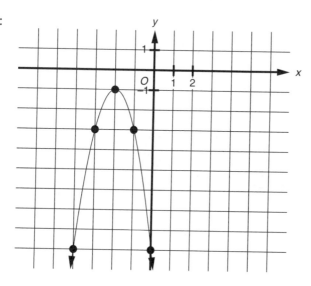

5.   Check that the graph matches the equation.
Is it $\smile$-shaped? ✓
Does it open in the right direction? $a < 0$, should open down. ✓
Does it pass through the y-intercept, $(0, -9)$? ✓

- -

Now it is time for you to try.

Problem 1

Graph: $y = -x^2 + 4x + 5$

- -

Solution: 1.   The equation is already in standard form.

2.   The axis of symmetry is

3.   Make a table of values:

x	$y = -x^2 + 4x + 5$	(x, y)

4.   Graph the points and connect them with a smooth curve.

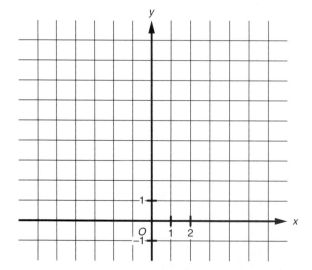

5. Check

Is it $\smile$-shaped?

a _____; graph should open _____.

Does it pass through the y–intercept, _____?

Problem 2

Graph: $y = x^2 + 2$.

Solution:

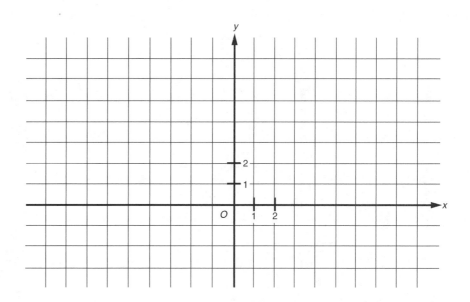

In the examples so far, our tables have included seven consecutive integer x-values. This works well for most parabolas that you might reasonably expect to graph by hand (without technology). If you are graphing a parabola where a is a fraction, you may want to pick your x-values to minimize the number of fractions in your calculations. The following example illustrates this.

EXAMPLE 4

Graph: $y = \dfrac{1}{2}x^2 - 3x - 1$

Solution:
1. The equation is already in standard form.

2. The axis of symmetry is $x = -\dfrac{b}{2a} = -\dfrac{(-3)}{2\left(\dfrac{1}{2}\right)} = 3$, or $x = 3$.

3. Make a table of values including the axis of symmetry and three integral x-values on each side. It is desirable to include the axis of symmetry in the table, so we will keep $x = 3$. But because $a = \dfrac{1}{2}$, choosing even numbers for the remaining x-values will lead to integral y-values, which are more convenient for graphing.

x	$y = \dfrac{1}{2}x^2 - 3x - 1$	(x, y)
-2	$y = \dfrac{1}{2}(-2)^2 - 3(-2) - 1$ $= 7$	$(-2, 7)$
0	$y = \dfrac{1}{2}(0)^2 - 3(0) - 1$ $= -1$	$(0, -1)$
2	$y = \dfrac{1}{2}(2)^2 - 3(2) - 1$ $= -5$	$(2, -5)$
3	$y = \dfrac{1}{2}(3)^2 - 3(3) - 1$ $= -5.5$	$(3, -5.5)$
4	$y = \dfrac{1}{2}(4)^2 - 3(4) - 1$ $= -5$	$(4, -5)$
6	$y = \dfrac{1}{2}(6)^2 - 3(6) - 1$ $= -1$	$(6, -1)$
8	$y = \dfrac{1}{2}(8)^2 - 3(8) - 1$ $= 7$	$(8, 7)$

4. Plot the points and connect them with a smooth curve.

Answer:

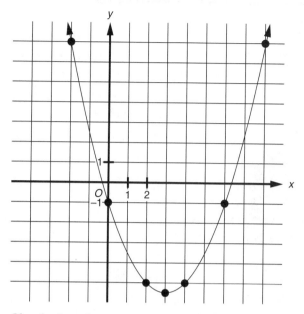

5. Check that the graph matches the equation.
    Is it ⌣-shaped? ✓
    Does it open in the right direction? $a > 0$, should open up. ✓
    Does it pass through the y-intercept, $(0, -1)$? ✓

ALGEBRA AND THE CALCULATOR (Optional)

Graphing Quadratic Equations

The procedure for graphing a quadratic equation using a calculator is basically the same as for a linear equation. Finding the vertex can be done several different ways.

EXAMPLE 5

Graph: $y = x^2 - 4x - 1$

Solution: Type the equation into Y= to find the table of values and see the graph.
Use the following keystrokes:

Y= CLEAR X,T,θ,n x^2 − 4 X,T,θ,n − 1 ENTER .

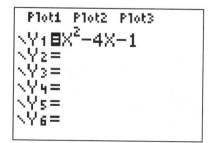

To see the y–values, press 2nd GRAPH .

X	Y₁	
0	-1	
1	-4	
2	-5	
3	-4	
4	-1	
5	4	
6	11	
X=0		

If the axis of symmetry is an integer value, you can see a symmetric pattern in the y-values in the table. Look at the table above; see how -1 and -4 are repeated around -5. Scroll the table so that the row with $x = 2$ and $y = -5$ is in the middle of the table.

X	Y₁	
-1	4	
0	-1	
1	-4	
2	-5	
3	-4	
4	-1	
5	4	
X=-1		

Use these seven points to plot your graph. When you are done, check your graph by comparing it to the calculator's graph. The standard window ZOOM 6 is shown below. The parabola looks like it has a horizontal line at the bottom, but it does not.

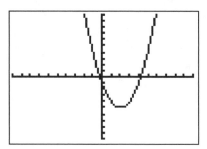

Set the values in the window to match the values in the table, then press GRAPH for a more accurate graph.

```
WINDOW
  Xmin=-1
  Xmax=5
  Xscl=1
  Ymin=-5
  Ymax=4
  Yscl=1
↓Xres=1
```

Answer: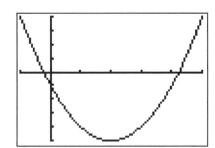

===

EXAMPLE 6

Graph: $y = 4x^2 - 12x + 3$

Solution: Type the equation into Y= to find the table of values and see the graph. Use the following keystrokes:

Y= CLEAR 4 X,T,θ,n x^2 – 1 2 X,T,θ,n + 3 ENTER.

```
Plot1 Plot2 Plot3
\Y1䊐4X²-12X+3
\Y2=
\Y3=
\Y4=
\Y5=
\Y6=
```

To see the *y*-values, press 2nd GRAPH.

```
   X   | Y1
  -1   | 19
   0   | 3
   1   | -5
   2   | -5
   3   | 3
   4   | 19
        | 43
  X=5
```

The axis of symmetry is not an integer, so use the formula for the axis of symmetry, $x = \dfrac{-b}{2a}$, to find the value. For $y = 4x^2 - 12x + 3$, $a = 4$ and

$b = -12$. To evaluate $x = \dfrac{-(-12)}{2(4)} = \dfrac{3}{2} = 1.5$ on your calculator, press 2nd

MODE CLEAR ALPHA Y= 1 (–) ((–) 1 2) > 2 (4) > ENTER
MATH 2 ENTER.

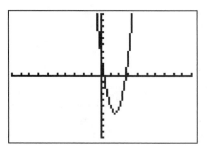

Look at the calculator's graph. The standard window ZOOM 6 is shown below.

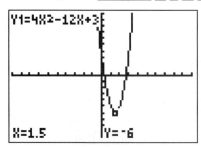

You need to graph points at $x = 1.5$ and at least two points on each side. To find the vertex, $(1.5, -6)$, press TRACE 1 . 5 ENTER.

Use the table above for $x = 0$, 1, 2, and 3. The y-values at $x = -1$ and 4 are too large to plot. Match the window to your table of values for x and y.

Answer:

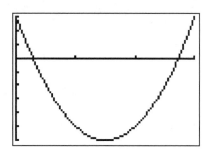

EXAMPLE 7

Graph: $y = -4x^2 + 6x + 3$

      a.   Find the coordinates of the vertex.

      b.   Find the x-intercepts.

Solution: Use the standard viewing window $\boxed{\text{ZOOM}}$ $\boxed{6}$.

Create a table of values for the equation with the following keystrokes:

$\boxed{\text{Y=}}$ $\boxed{\text{CLEAR}}$ $\boxed{\text{(–)}}$ $\boxed{4}$ $\boxed{\text{X,T,θ,n}}$ $\boxed{x^2}$ $\boxed{+}$ $\boxed{6}$ $\boxed{\text{X,T,θ,n}}$ $\boxed{+}$ $\boxed{3}$ $\boxed{\text{ENTER}}$.

To see the y-values, press $\boxed{\text{2nd}}$ $\boxed{\text{TABLE}}$.
To see the graph, press $\boxed{\text{GRAPH}}$.

    a.   Return to the table of y-values; press $\boxed{\text{2nd}}$ $\boxed{\text{TABLE}}$.
        This is a quadratic equation with $a = -4$, $b = 6$, and $c = 3$.

        The x-coordinate of the vertex is given by the formula $x = \dfrac{-6}{2(-4)}$.

        Using the formula, key in the x-coordinate of the vertex as follows:

        $\boxed{\text{ALPHA}}$ $\boxed{\text{Y=}}$ $\boxed{\text{ENTER}}$ $\boxed{\text{(–)}}$ $\boxed{6}$ $\boxed{>}$ $\boxed{2}$ $\boxed{(}$ $\boxed{(}$ $\boxed{\text{(–)}}$ $\boxed{4}$ $\boxed{)}$ $\boxed{>}$ $\boxed{\text{ENTER}}$ $\boxed{\text{MATH}}$ $\boxed{2}$ $\boxed{\text{ENTER}}$.

Answer: The vertex is located at the point (0.75, 5.25).

As a check, use $\boxed{\text{TRACE}}$ $\boxed{0.75}$ $\boxed{\text{ENTER}}$ to verify that the answer is reasonable.

    b.   To find the negative x-intercept, also called a zero, the steps are the same as those introduced in Unit 23.

        Press $\boxed{\text{2nd}}$ $\boxed{\text{CALC}}$ $\boxed{2}$.

        At the <u>Left Bound?</u> prompt, move the cursor to a point just to the left of the x-intercept and press $\boxed{\text{ENTER}}$.

        At the <u>Right Bound?</u> prompt, move the cursor to a point just to the right of the x-intercept and press $\boxed{\text{ENTER}}$.

        At the <u>Guess?</u> prompt, move the cursor between the boundaries and press $\boxed{\text{ENTER}}$.

Answer: The negative x-intercept is –0.396.
        For practice, you find the positive x-intercept.

Answer: The positive x-intercept is 1.896.

To summarize, in Unit 25 we defined (and you should now be able to identify) a second-degree equation in two variables, of the type $y = ax^2 + bx + c$ with $a \neq 0$. Also, you should be able to graph such an equation, using a minimum number of well-chosen points, based on your knowledge of quadratic equations.

Recall that, in short, the graphing approach is as follows:

1. Write the equation in standard form.

2. Find the x-value of the axis of symmetry: $x = -\dfrac{b}{2a}$.

3. Make a table of values including the axis of symmetry and three integral x-values on each side of the axis, a total of seven points. (If the y-values become very large, we may settle for fewer x-values.)

4. Plot the points and connect them with a smooth curve.

5. Check that your graph matches the equation.

- Is it $\smile$-shaped (not a line)?

- Does it open in the right direction, up if $a > 0$, down if $a < 0$?

- Does it pass through the y-intercept, $(0, c)$?

Now try the following exercises.

EXERCISES

Graph each of the following. Optional—verify the reasonableness of your graph by using a graphing calculator.

1. $y = x^2 + 6x + 8$

2. $y = x^2 + 2x - 8$

3. $y = -\dfrac{1}{2}x^2$

4. $y = -x^2 - 4x - 3$

5. $y = x^2 - 6x + 5$

6. $y = 4 + 2x^2$

7. $y = x^2 - 4x + 4$

8. $y = 9 - x^2$

9. $y = x^2 + 6x + 9$

10. $y = 2x^2 + 4x - 1$

11. $y = 5x^2 - 20x + 11$

12. $y = -x^2 + 6x$

13. $y = 3x^2 - 3x + 2$

14. $y = 2x^2 - 12x + 3$

15. $y = \dfrac{-1}{4}x^2 + 3x - 8$

Solutions to Problems

1.

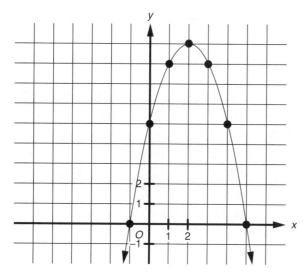

2.

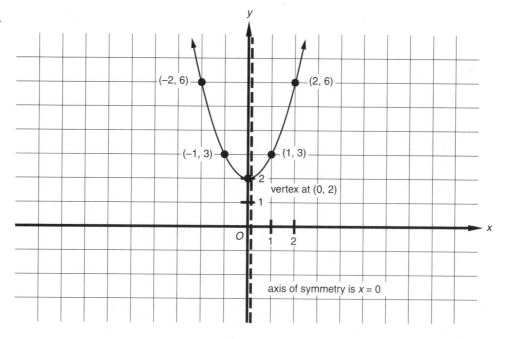

UNIT 25

Functions

This unit introduces the concept of a function. When you have completed the unit, you will be able to read and to use functional notation and identify the domain.

We will start with a diagram of a machine.

We can define our machine to do almost anything. Since this is a book about algebra, our machine is one that does calculations with numbers. Suppose that the numbers 5, 7, 8, and 9 are located as shown and that the machine "takes a number and adds 3."

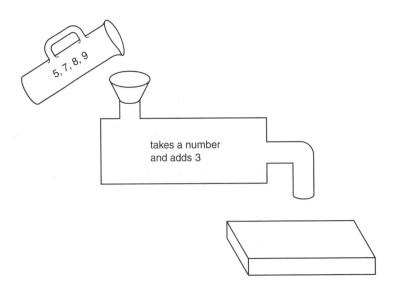

It should be obvious what happens as each number is put into the machine.

If 5 is entered, the machine takes the number and adds 3, and out comes 8.

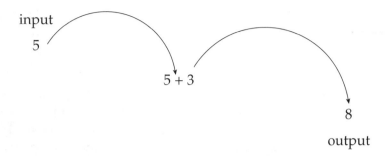

input

5

5 + 3

8

output

In a similar fashion, if 7 had been put in, the output would have been 10. If 8 had been put in, the output would have been 11. If 9 had been put in, the output would have been 12. Write the numbers 8, 10, 11, and 12 in the box in the machine diagram.

The values that can be put into the machine are referred to as the domain. The **domain** is the set of all possible values that can be used as input. In this example, the domain is the set of numbers 5, 7, 8, and 9, or $D = \{5, 7, 8, 9\}$. Only these four numbers can be put into this particular machine. Why? Because that is the way it was created.

The values that come out are referred to as the range. The **range** is the set of all possible values that are output. In the diagram the range is the set of numbers 8, 10, 11, and 12, or $R = \{8, 10, 11, 12\}$.

The description "takes the number and adds 3" is referred to as the **rule**.

Notice that our diagram has three parts: the domain, the range, and the rule.

The domain, often denoted by x, is the **independent variable**, whereas the range, often denoted as y, is the **dependent variable**. A way to remember the distinction between the two is that we can select any value from the domain to put into the machine, but once a specific value of x is selected, such as 7, the output, or the y-value, is dependent on it.

By using x to denote the independent variable, it is possible to rewrite the rule algebraically as $x + 3$. At the same time, we will give the machine the name f. Typically f is the most common name used, although g and h are popular too.

The diagram has been drawn again to include all the terminology.

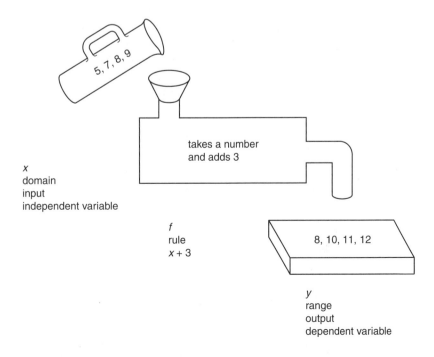

x
domain
input
independent variable

f
rule
$x + 3$

y
range
output
dependent variable

> Definition: A **function** is a rule that assigns to each element in the domain one and only one element in the range.

Like the diagram, the definition refers to three parts: domain, range, and rule. The most important part of the definition, though, is the *one and only one* phrase. For each value put in, only a single value can come out if the expression is to be a function. Otherwise the expression is classified as a **relation**. Functions will be the major emphasis in this unit.

Obviously we don't want to have to draw a diagram every time we talk about a function. All of the above information can be combined into a single statement using functional notation.

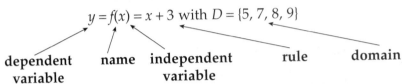

$$y = f(x) = x + 3 \text{ with } D = \{5, 7, 8, 9\}$$

| dependent variable | name | independent variable | rule | domain |

The above statement is read "*y* equals *f* of *x* equals *x* plus 3, with domain *D* equal to the set of numbers 5, 7, 8, and 9.

Caution: $f(x)$ is read "*f* of *x*" and does not mean multiplication. It is used to indicate that the name of the function is *f* and that the variable inside the parentheses, in this example, *x*, is the independent variable.

It is time to stop and try some questions. Using *f* as defined by the diagram, answer the following questions.

1. If $x = 7$, what is *y*?   Answer: 10
2. If $x = 9$, what is *y*?   Answer: 12
3. If $y = 11$, what is *x*?   Answer: 8

We're confident that you were able to answer each of these questions correctly without any difficulty whatsoever. The difficulty occurs when the same questions are asked using functional notation. Here are some examples.

EXAMPLE 1

Continue to use *f* as defined in the diagram:

a.   Find $f(7)$.   c.   Solve $f(x) = 8$ for *x*.

b.   Find $f(9)$.   d.   Find $f^{-1}(11)$.

Solution: a. $f(7)$, read "*f* of 7," is asking, if $x = 7$, what is *y*? It is exactly the same as question 1 above, and the answer is 10, or $f(7) = 10$.

b. $f(9)$ is asking the same thing as question 2 above but uses functional notation. The question is, if $x = 9$ and the rule is *f*, what is *y*? As in question 2 above, the answer is 12, or $f(9) = 12$.

c.   Solve $f(x) = 8$ for *x* means substitute 8 for $f(x)$ in the equation $f(x) = x + 3$, then solve for *x*: $8 = x + 3; x = 5$.

d. $f^{-1}(11)$ is read "*f* inverse of 11." The question is, if the *y*-value is 11, what is *x*? As in question 3 above, the answer is 8, or $f^{-1}(11) = 8$.

Here are a few problems for you to try. Use f as defined in the diagram to answer each of the following.

Problem 1

Find $f(5)$, $f^{-1}(10)$, and $f(8)$.

Solution:

Problem 2

Find $f(6)$. Be careful; the answer is not 9.

Solution:

EXAMPLE 2

Given: $y = g(x) = 2x + 5$ with the domain the set of real numbers.

a.   Identify each part of the expression.
b.   Find $g(3)$
c.   Find $g(-1)$
d.   Find $g(0)$
e.   Find $g^{-1}(13)$
f.   Find $g(a)$
g.   Find $g(x^2)$
h.   Find $g(x + 3)$
i.   Solve $g(x) = 17$ for x.

Solution: a. y is the dependent variable.

x is the independent variable.

g is the name of the function.

The rule is 2 times the number plus 5.

$g(x) = 2x + 5$
b. $g(3) = 2(3) + 5 = 6 + 5 = 11$
c. $g(-1) = 2(-1) + 5 = -2 + 5 = 3$
d. $g(0) = 2(0) + 5 = 0 + 5 = 5$

e. $g^{-1}(13)$

$$13 = 2x + 5$$
$$8 = 2x$$
$$4 = x$$
$$g^{-1}(13) = 4$$

f. $g(a) = 2(a) + 5 = 2a + 5$

g. $g(x^2) = 2(x^2) + 5 = 2x^2 + 5$

h. $g(x + 3) = 2(x + 3) + 5$
$$= 2x + 6 + 5$$
$$= 2x + 11$$

i. Solve $g(x) = 17$ for x: $17 = 2x + 5$
$$12 = 2x$$
$$6 = x$$

Problem 3

Given: $y = h(x) = 3x - 2$, with the domain the set of real numbers.

Find the following:

a. $h(4)$

b. $h(-5)$

c. $h^{-1}(13)$

d. $h(c)$

e. $h(y^2)$

f. Solve $h(x) = -8$ for x.

Although functions are commonly expressed in algebraic form, sets and graphs also can be used, as the next examples illustrate.

EXAMPLE 3

Given: $F = \{(1, 2), (3, 7), (3, 15)\}$.

Find: a. The domain.

b. The range.

c. $F(1)$.

d. Solve $F(x) = 15$ for x.

e. Is F a function?

Solution: F is the name. The ordered pair (1, 2) means that, when $x = 1$, $y = 2$. Similarly, (3, 7) means that when $x = 3$, $y = 7$, and (3, 15) means that when $x = 3$, $y = 15$. In this example the actual rule is not given.

    a. The domain is the set of all x-values: $D = \{1, 3\}$. Note: There is no need to write the 3 twice.

    b. The range is the set of y-values: $R = \{2, 7, 15\}$.

    c. $F(1) = 2$. The notation is asking, when $x = 1$, what is y?

    d. The solution to $F(x) = 15$ is $x = 3$. The notation is asking, when $y = 15$, what is x?

    e. No, F is not a function. When $x = 3$, there are two different values for y, 7 and 15.

EXAMPLE 4

Does the graph at the right specify a function?

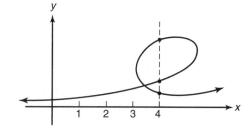

Solution: No, for example, when $x = 4$, there are three values for y.

Notice that if you are able to draw a vertical line that intersects a graph at more than one point, the graph does not represent a function. This is referred to as the **vertical line test**.

When a function is expressed in algebraic form, the domain is seldom mentioned. Unless specified, the **domain** of a function is the set of all values for which the dependent variable is defined and real. This is usually the set of real numbers. Remember, the real numbers are all the counting numbers, zero, whole numbers, positive and negative integers, fractions, and rational and irrational numbers. There are two important exceptions for the domain.

1. Division by zero is undefined. Any values of the variable that would make a fraction undefined are excluded from the domain.

2. The square root (or any even root) of a negative number is not real. Any expression in a square root will be non-negative.

EXAMPLE 5

Find the domain of $y = f(x) = \dfrac{x + 3}{2x - 1}$.

Solution: The denominator of the fraction cannot equal 0. Set the denominator equal to 0 and solve for x. Then state that this x-value cannot be in the domain.

$$2x - 1 = 0$$
$$2x = 1$$
$$x = \frac{1}{2}$$

The domain is the set of all real numbers except for $x = \dfrac{1}{2}$, sometimes written as all real numbers with $x \neq \dfrac{1}{2}$.

EXAMPLE 6

Find the domain of $y = g(x) = \sqrt{x + 4}$.

Solution: The expression inside the square root symbol must be non-negative. Set the expression to be greater than or equal to 0 and solve for x. These values of x are the domain.

$$x + 4 \geq 0$$
$$x \geq -4$$

The domain is all real numbers greater than or equal to -4, sometimes written as $x \geq -4$.

ALGEBRA AND THE CALCULATOR (Optional)

Creating a Table of Values for a Function

In previous units we created tables of values for linear equations and quadratic equations. In this section the notation and terminology have been changed to create a table of values for a function, but the steps remain identical.

EXAMPLE 7

Given: $y = f(x) = 3.2x + 2.1$.

Find: $f(14)$, $f(7.5)$, $f(1.45)$, $f(0)$, $f(-0.023)$, $f(-5)$, and $f(-8.1)$.

Solution: This example is similar to the ones we did in Unit 23 except that now we are using functional notation and our example is called a linear function.

To enter the function, press the following keys.

[Y=] [CLEAR] [3] [.] [2] [X,T,θ,n] [+] [2] [.] [1] [ENTER] .

You should see the following:

```
Plot1  Plot2  Plot3
\Y1◻3.2X+2.1
\Y2=
\Y3=
\Y4=
\Y5=
\Y6=
\Y7=
```

To find the desired y-values, set the table to Ask (Unit 21), then press [2nd] [TABLE] .

To find $f(14)$, press [1] [4] [ENTER] .

To find $f(7.5)$, press [7] [.] [5] [ENTER] .

Repeat, entering the remaining values: 1.45, 0, −0.023, −5, and −8.1.

You should see the following:

```
X         Y1
14        46.9
7.5       26.1
1.45      6.74
0         2.1
-.023     2.0264
-5        -13.9
-8.1      -23.82
X=-8.1
```

Confirm that the function's graph is a line by viewing its graph.

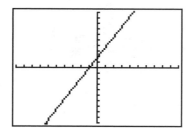

You should now understand the basic concept of a function along with the terms *domain*, *range*, and *rule*. You should also be able to read and evaluate questions written in functional notation.

Before beginning the next unit you should do all the following exercises.

EXERCISES

Given: $f(x) = x - 2$, find each of the following:

1. $f(7)$

2. $f(-3)$

3. $f(3.7)$

4. $f(0)$

Given: $g(x) = 3x$, find each of the following:

5. $g(2)$

6. $g(-5)$

7. $g(0)$

8. $g(-4.1)$

    a. $g(x - 5)$

    b. $g(a^2)$

Given: $h(x) = 5x + 1$, find each of the following:

9. $h(10)$

10. $h(3)$

11. $h(0)$

12. $h(-4)$

    a. $h(k)$

    b. $h(-x)$

    c. Solve $h(x) = -14$ for x.

Given: $f(x) = 1 - x^2$, find each of the following:

13. $f(3)$

14. $f(-5)$

15. $f(-1)$

Given: $G = \{(5, 6), (7, 7), (8, 5), (6, 11)\}$

16. State the domain for G.

17. State the range for G.

18. Find $G(8)$.

19. Find $G(5)$.

20. Find $G^{-1}(11)$.

21. Find $G^{-1}(5)$.

22. Is G a function?

C23. Given the function $y = f(x) = 2x^2 - 1.5x + 2$, find $f(2.5)$ and $f(0.25)$.

C24. Given the function $y = f(x) = 3x^4 + 2x^3 - x^2 + 1$, find $f(2)$ and $f(1.5)$.

Solutions to Problems

1. $f(5) = 5 + 3 = 8$

   $f^{-1}(10) \rightarrow 10 = x + 3 \rightarrow x = 7 \rightarrow f^{-1}(10) = 7$

   $f(8) = 8 + 3 = 11$

2. $f(6)$ is undefined because 6 is not an element of the domain. f is defined only for the numbers 5, 7, 8, and 9, which was an arbitrary decision on our part; but that was the way we defined the machine.

3. a. $h(4) = 3(4) - 2 = 12 - 2 = 10$

   b. $h(-5) = 3(-5) - 2 = -15 - 2 = -17$

   c. $h^{-1}(13)$

   $$13 = 3x - 2$$
   $$15 = 3x$$
   $$5 = x$$
   $$h^{-1}(13) = 5$$

   d. $h(c) = 3(c) - 2 = 3c - 2$

   e. $h(y^2) = 3(y^2) - 2 = 3y^2 - 2$

   f. Solve $h(x) = -8$ for x. $-8 = 3x - 2$

   $$-6 = 3x$$
   $$-2 = x$$

UNIT 26

Solving Systems of Equations

The purpose of this unit is to provide you with an understanding of systems of equations. When you have finished the unit, you will be able to solve systems of two linear equations in two variables.

Recall the following definition from Unit 23:

> Definition: An equation that has the form $ax + by = c$, with a, b, and c being real numbers, a and b not both zero, is a **linear equation in two variables**.

By a "solution to an equation in two variables" we mean the ordered pairs of values of x and y that satisfy the equation. The procedure outlined earlier for locating points on the graph of an equation also yields solutions to the equation. In fact, solutions are often written as ordered pairs.

There are infinitely many solutions to an equation in two variables.

EXAMPLE 1

Solve: $x + y = 8$.

Solution: Recall how to locate a point:

1. Select a convenient value for x.   Let $x = 2$.

2. Substitute into the equation. $(2) + y = 8$

3. Solve for y. $y = 6$

Then (2, 6) is a point on the graph of $x + y = 8$

and

$x = 2$ and $y = 6$ is called a solution to the equation $x + y = 8$.

Without going through the calculations, (3, 5) is a point on the graph of $x + y = 8$

and

$x = 3$ and $y = 5$ is another solution to the equation.

There are infinitely many points on the graph

and

there are infinitely many solutions to the equation, some of which are

$$
\begin{array}{llll}
x = 1 & \text{and} & y = 7 & \text{or simply } (1, 7) \\
x = 0 & \text{and} & y = 8 & \text{or} \qquad (0, 8) \\
x = 2.5 & \text{and} & y = 5.5 & \text{or} \qquad (2.5, 5.5) \\
x = -3 & \text{and} & y = 11 & \text{or} \qquad (-3, 11)
\end{array}
$$

A system of equations means that there is more than one equation.

Definition: **The solutions** to a system of equations are the pairs of values of x and y that satisfy *all* the equations in the system.

In this unit we will deal only with linear equations in two variables, saving quadratic equations for the next unit.

In general, the number of solutions to any system of *linear* equations is either one, none, or infinitely many.

$$
\begin{cases}
2x + y = 24 \\
x - y = 6
\end{cases}
$$

is an example of a system of linear equations. The solution to this system is $x = 10$ and $y = 4$, or simply (10, 4), because this pair of values satisfies *both* equations in the system.

To verify, check by substitution.

$$
\begin{array}{ll}
2x + y = 24 & x - y = 6 \\
2(10) + 4 \overset{?}{=} 24 & 10 - 4 \overset{?}{=} 6 \\
24 = 24 & 6 = 6
\end{array}
$$

The rest of the unit deals with finding such solutions.

Systems of two equations in two variables can be solved graphically or algebraically. We will first look at the graphical method.

SOLVING A SYSTEM GRAPHICALLY

To solve a system of two linear equations in two variables graphically:

1. Graph both lines on the same set of axes. Label each line.

2. Write the solution, the point where the two lines intersect. (If the two lines do not intersect, the system has no solution.)

EXAMPLE 2

Solve graphically and check: $\begin{cases} 3x - y = -7 \\ 5x + 5 = -5x \end{cases}$

- -

Solution: 1. We need to graph both lines. You can use any method that seems convenient; we'll use the slope-intercept method, which means we first need to solve both equations for y.

$$
\begin{array}{ll}
3x - y = -7 & 5y + 5 = -5x \\
-y = -3x - 7 & 5y = -5x - 5 \\
y = 3x + 7 & y = -x - 1
\end{array}
$$

For the first line, we have a y-intercept of 7 and a slope of $\dfrac{3}{1}$.

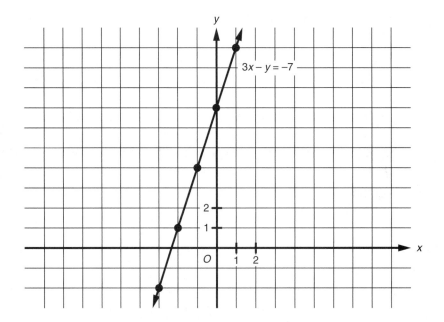

For the second line, the y-intercept is -1 and the slope is $\dfrac{-1}{1}$.

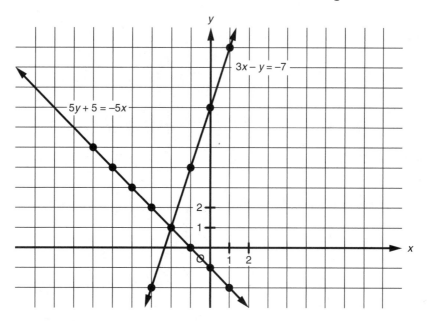

The solution is the point where the two lines intersect. In this problem, the solution is $(-2, 1)$, which means that $x = -2$ and $y = 1$.

To check that this really is the solution to the system, we need to show the point $(-2, 1)$ works in both equations. Do this, one equation at time, by substituting and evaluating. Remember, always check in the original equations.

Check:
$$3x - y = -7 \qquad\qquad 5y + 5 = -5x$$
$$3(-2) - 1 = -7 \qquad\qquad 5(1) + 5 = -5(-2)$$
$$-7 = -7 \checkmark \qquad\qquad 10 = 10 \checkmark$$

This is indeed the solution.

Answer: $x = -2$ and $y = 1$, which may be written more compactly as $(-2, 1)$.

Try one yourself.

Problem 1

Solve graphically and check: $\begin{cases} 4x - y - 1 = 0 \\ 2x = 17 - y \end{cases}$

Hint: Because the second equation has a relatively large y-intercept, a table of values will be more convenient than the slope-intercept method for that equation.

Solution:

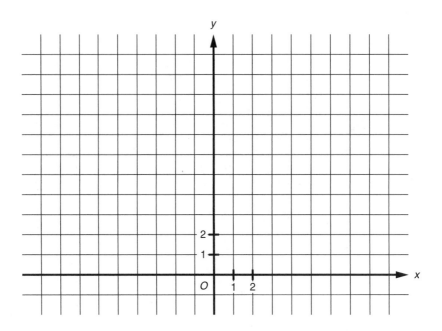

Now let's look at solving a system of two linear equations in two unknowns algebraically. We already know how to solve a linear equation in just one variable. Our strategy for a system of equations will be to simplify the system to one equation in just one variable. This is called **elimination** because we (temporarily) eliminate one of the variables from the problem. There are two common methods for elimination—the **addition method** and the **substitution method**.

ELIMINATION BY ADDITION

A linear equation written in the form $ax + by = c$ is said to be in **standard form**. It is advisable to simplify and write both equations in standard form before attempting to solve a system of linear equations by elimination by addition.

The procedure I will use involves four steps:

1. **Multiply (if necessary)** the equations by constants so that the coefficients of the x or the y variable are the opposites of one another.

2. **Add** the equations from step 1.

3. **Solve** the equation from step 2.

4. **Substitute** the answer from step 3 back into one of the original equations, and solve for the second variable.

Here are some examples that illustrate the use of these steps in solving a system of linear equations.

EXAMPLE 3

Solve: $\begin{cases} x + y = 7 \\ 5x - 3y = 11 \end{cases}$.

Solution: Multiply the first equation by 3 so that the y-coefficients are the opposites of one another.

    1. **Multiply.** $3(x + y = 7)$

                         $5x - 3y = 11$

    2. **Add.** $\quad\quad 3x + 3y = 21$

                         $\underline{5x - 3y = 11}$

    3. **Solve.** $\quad\quad 8x \quad\quad = 32$

                         $x \quad\quad = 4$

    4. **Substitute back** into one of the original equations. I usually try to select the simpler of the two original equations for the substitution.

        If $x = 4$ and $x + y = 7$,

                    $(4) + y = 7$,

                         $y = 3$.

Answer: Solution to system is $x = 4$ and $y = 3$ or, written another way, $(4, 3)$.

    Note: Back in step 1 we could just as well have multiplied the equation by -5 so that the x-coefficients would have been the opposites of one another.

EXAMPLE 4

Solve: $\begin{cases} 3x + 2y = 12 \\ y = 2x - 1 \end{cases}$.

Solution: Write the equations in standard form.

$$3x + 2y = 12$$
$$-2x + y = -1$$

Multiply the second equation by -2 so that the y-coefficients are the negatives of one another.

    1. **Multiply.** $\quad 3x + 2y = 12$

                      $-2(-2x + y = -1)$

    2. **Add.** $\quad\quad\quad 3x + 2y = 12$

                     $\underline{4x - 2y = \ 2}$

    3. **Solve.** $\quad\quad\quad 7x \quad\quad = 14$

                      $x \quad\quad = 2$

4. **Substitute back** into an original equation.

If $x = 2$ and $y = 2x - 1$,

$$y = 2(2) - 1,$$
$$y = 3.$$

Answer: Solution to system is $x = 2$ and $y = 3$; $(2, 3)$.

EXAMPLE 5

Solve: $\begin{cases} 3x - y = -7 \\ 5y + 5 = -5x \end{cases}$.

Solution: Write the second equation in standard form before starting the procedure.

Rewrite. $5x + 5y = -5$

The equations are now: $3x - y = -7$

$$5x + 5y = -5$$

1. **Multiply.** $5(3x - y = -7)$

2. **Add.** $15x - 5y = -35$

$$\underline{5x + 5y = \ -5}$$

3. **Solve.** $20x \qquad = -40$

$$x \qquad = -2$$

4. **Substitute back** into an original equation.

If $x = -2$ and $5y + 5 = -5x$

$$5y + 5 = -5(-2)$$
$$5y + 5 = 10$$
$$5y = 5$$
$$y = 1.$$

Answer: Solution to system is $x = -2$ and $y = 1$; $(-2, 1)$.

From an algebraic point of view,

$x = -2$ and $y = 1$ is the solution to the system of equations $3x - y = -7$ and $5y + 5 = -5x$.

From a graphical point of view,

$(-2, 1)$ is the point of intersection for two lines whose equations are given by $3x - y = -7$ and $5y + 5 = -5x$ as shown in Example 2.

Now it is your turn to solve a system of equations.

Problem 2

Solve using elimination: $\begin{cases} 4x - y - 1 = 0 \\ 2x = 17 - y \end{cases}$.

Solution:

Hint: First write equations in standard form.

1. **Multiply**, if necessary.

2. **Add.**

3. **Solve.**

4. **Substitute back** into an original equation.

Problem 3

Solve algebraically: $\begin{cases} x + 2y = 1 \\ 5x + 3y = 26 \end{cases}$.

Solution: 1. **Multiply.**

2. **Add.**

3. **Solve.**

4. **Substitute back** into an original equation.

Many times it is necessary to multiply the two equations by different constants in order to make the coefficients of the x or the y variable be the negatives of one another.

EXAMPLE 6

Solve: $\begin{cases} 2x + 3y = 9 \\ 5x + 2y = 17 \end{cases}$.

Solution: 1.   **Multiply.** $-5(2x + 3y = 9)$

$2(5x + 2y = 17)$

2.   **Add.** $-10x - 15y = -45$

$\underline{10x\ +4y = 34}$

3.   **Solve.** $-11y = -11$

$y = 1$

4.   **Substitute back** into an original equation.

If $y = 1$ and $2x + 3y = 9$,

$2x + 3(1) = 9$,

$2x = 6$,

$x = 3$.

Answer: Solution to system is $x = 3$ and $y = 1$.

Did you notice that there were several possibilities for the constants used in step 1? We could just as well have multiplied the first equation by 2 and the second equation by -3.

General Rule: Multiply by the coefficient of the other equation. One of the multipliers must be the opposite of the other coefficient.

Here are two problems for you to solve on your own.

Problem 4

Solve: $\begin{cases} 3x - 2y = 5 \\ -4x + 3y = 1 \end{cases}$.

Solution:

As you probably suspected, not all systems have nice integral values as solutions. Be prepared for fractions.

Problem 5

Solve: $\begin{cases} \dfrac{x}{3} + y = 3 \\ 4x + 3y = 0 \end{cases}$.

Solution:

Hint: Simplify the first equation by clearing the fractions before attempting to solve this system of linear equations.

Recall that at the beginning of this unit it was stated that the number of solutions to a system of linear equations will be one, none, or infinitely many. Thus far, we have seen only systems with one solution. Examples 7 and 8 will illustrate the other two situations.

EXAMPLE 7

Solve: $\begin{cases} 3x + 4y = 2 \\ 2y = 4 - \dfrac{3}{2}x \end{cases}$.

Solution: First simplify the second equation, and write it in standard form.

$$2\left(2y = 4 - \frac{3}{2}x \right)$$

$$4y = 8 - 3x$$

$$3x + 4y = 8$$

1. **Multiply.** $-1(3x + 4y = 2)$

   $3x + 4y = 8$

2. **Add.** $-3x - 4y = -2$

   $\underline{3x + 4y = 8}$

   $0 = 6$

But $0 = 6$ is a false statement; and because $0 = 6$ is a false statement, there are no values of x and y that would ever satisfy this equation.

Answer: There is no solution to the system.

The equations are said to be **inconsistent** when there is no solution to the system.

Graphically, no solution to a system of two linear equations means that there is no point of intersection for the two lines. The lines are parallel.

We will verify, by graphing, that the system in Example 7 represents two parallel lines.

Problem 6

Solve graphically: $\begin{cases} 3x + 4y = 2 \\ 2y = 4 - \dfrac{3}{2}x \end{cases}$.

Solution:

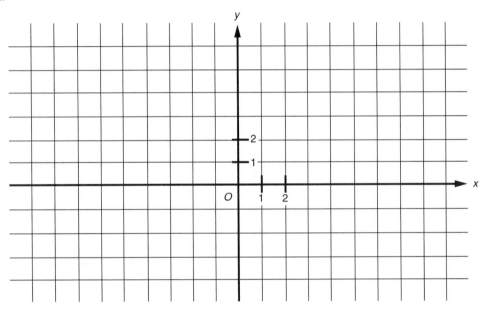

EXAMPLE 8

Solve: $\begin{cases} 3x - y = 5 \\ 6x - 2y - 10 = 0 \end{cases}$.

Solution: $-2(3x - y = 5)$

$6x - 2y = 10$

$-6x + 2y = -10$

$\underline{6x - 2y = 10}$

$\qquad 0 = 0 \quad$ is a true statement.

This time our procedure results in a true statement or, as defined in an earlier unit, an identity.

Answer: There are infinitely many solutions to the system.

Every solution to the first equation will satisfy the second equation as well. A closer inspection of the two equations should reveal why this is true; the two equations are identical.

If x is any real number, then $y = 3x - 5$.

The equations are said to be **dependent** when there are infinitely many solutions to the system.

Graphically, "many solutions to a system of two linear equations" means there are infinitely many points of intersection for the two lines. The lines coincide.

Let's review what has just been stated.

No solution to a system is indicated when the elimination method leads to a false statement, such as $0 = 6$. The equations are said to be **inconsistent**. The lines are **parallel**.

Infinitely many solutions to a system are indicated when the elimination method leads to an identity, such as $0 = 0$. The equations are said to be **dependent**. The lines **coincide**.

Problem 7

Solve algebraically: $\begin{cases} 5x - y = 3 \\ 2y = 10x - 6 \end{cases}$.

Solution:

Problem 8

Solve algebraically: $\begin{cases} 4x + 3y = 9 \\ 2y = 4 - \dfrac{8}{3}x \end{cases}$.

Solution:

ELIMINATION BY SUBSTITUTION

Elimination by substitution is another method to solve systems of equations. If one of the variables has a coefficient of 1, it is sometimes more efficient to eliminate that variable by substitution rather than by multiplication and addition. Let us illustrate what we mean with the next example.

EXAMPLE 9

Solve: $\begin{cases} y = 2x - 8 \\ 3x + 2y = 12 \end{cases}$.

Solution: Use $y = 2x - 8$ to substitute into the second equation.

$$3x + \ \ 2y \ \ \ \ \ = 12$$
$$3x + 2(2x - 8) = 12 \qquad \text{Replace } y \text{ with } 2x - 8.$$
$$3x + \ \ 4x - 16 = 12$$
$$7x \ \ \ \ \ \ \ \ = 28$$
$$x \ \ \ \ \ = 4$$

Finish the problem as before by substituting back into an original equation to find y:

$$y \ \ \ \ \ = 0$$

Answer: Solution to system is $x = 4$ and $y = 0$; (4, 0).

We use elimination by substitution whenever possible, because it often requires less rewriting of the equations.

EXAMPLE 10

Solve: $\begin{cases} y = 3x + 1 \\ 3x + 4y = -26 \end{cases}$.

Solution: Use $y = 3x + 1$ to substitute into the second equation.

$$3x + 4y \qquad = -26$$
$$3x + 4(3x + 1) = -26 \qquad \text{Replace } y \text{ with } 3x + 1.$$
$$3x + 12x + 4 = -26$$
$$15x \qquad = -30$$
$$x \qquad = -2$$

and, by substituting back into an original equation,

$$y \qquad = -5$$

Answer: Solution to system is $x = -2$ and $y = -5$; $(-2, -5)$.

Let's do one more example, and then you can try some problems.

EXAMPLE 11

Solve: $\begin{cases} 5x - 2y = 11 \\ y = \dfrac{5}{2}x - 3 \end{cases}$.

Solution: Use $y = \dfrac{5}{2}x - 3$ to substitute into the first equation.

$$5x - 2y \qquad = 11$$
$$5x - 2\left(\frac{5}{2}x - 3\right) = 11 \qquad \text{Replace } y \text{ with } \frac{5}{2}x - 3.$$
$$5x - 5x + 6 = 11$$
$$6 = 11 \text{ is a false statement.}$$

Answer: There is no solution to the system.

Here are two problems for you to solve. The first one is easy; the second is similar to the examples.

Problem 9

Solve: $\begin{cases} 3x + 5y = 14 \\ y = -2 \end{cases}$.

Solution:

Problem 10

Solve by substitution: $\begin{cases} y = 7x + 2 \\ x - 3y = -6 \end{cases}$.

Solution:

ALGEBRA AND THE CALCULATOR (Optional)

Finding Points of Intersection

We will revisit Example 10, but this time we will find the solution using a graphing calculator.

EXAMPLE 12

Solve: $\begin{cases} y = 3x + 1 \\ 3x + 4y = -26 \end{cases}$.

Solution: Graph and find the point of intersection using a graphing calculator. First solve the second equation for y.

$$3x + 4y = -26$$
$$4y = -3x - 26$$
$$\frac{4y}{4} = \frac{-3x - 26}{4}$$
$$y = \frac{-3x - 26}{4}$$

Use the standard viewing window $\boxed{\text{ZOOM}}\ \boxed{6}$.

Key in *both* functions as follows:

$\boxed{\text{Y=}}$ $\boxed{3}$ $\boxed{\text{X,T,θ,n}}$ $\boxed{+}$ $\boxed{1}$ $\boxed{\text{ENTER}}$.

$\boxed{\text{Y=}}$ $\boxed{\text{ALPHA}}$ $\boxed{\text{Y=}}$ $\boxed{\text{ENTER}}$ $\boxed{(-)}$ $\boxed{3}$ $\boxed{\text{X,T,θ,n}}$ $\boxed{-}$ $\boxed{2}$ $\boxed{6}$ $\boxed{\vee}$ $\boxed{4}$.

and view the graph.

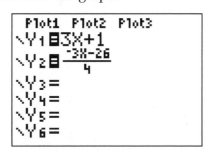 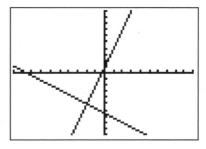

To find the point of intersection, press $\boxed{\text{2nd}}$ $\boxed{\text{CALC}}$ $\boxed{5}$.

At the <u>First curve?</u> prompt, press $\boxed{\text{ENTER}}$.

At the <u>Second curve?</u> prompt, press $\boxed{\text{ENTER}}$.

At the <u>Guess?</u> prompt, move the cursor to where the intersection appears to be, and press $\boxed{\text{ENTER}}$.

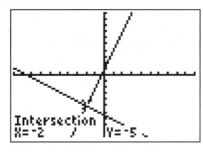

Answer: Intersection $X = -2$ $Y = -5$ appears at the bottom of the screen.

You should now be able to solve any system of two linear equations in two variables.

To solve graphically:

1. Graph each equation using the slope-intercept method or a table of values.

2. Label the equations.

3. Write the solution—the point of intersection.

To solve algebraically using elimination by addition, the four basic steps involved are:

1. **Multiply**, if necessary.
2. **Add.**
3. **Solve.**
4. **Substitute back.**

To solve algebraically using substitution:

1. Solve one equation for x or y.
2. Substitute the expression for x or y into the other equation.
3. Solve.
4. Substitute back.

Also remember that the number of solutions to any system of linear equations is either one, none, or infinitely many.

Before beginning the next unit you should do the following exercises.

EXERCISES

Solve these systems graphically.

1. $\begin{cases} y = 3x + 2 \\ x - 3y = -6 \end{cases}$

2. $\begin{cases} 2x + y = 8 \\ 2x - 3y = 0 \end{cases}$

3. $\begin{cases} y = 2x + 3 \\ y + 2 = 4x + 1 \end{cases}$

4. $\begin{cases} y = -\dfrac{1}{2}x - 4 \\ x + 2y = 6 \end{cases}$

Solve these systems using elimination by addition.

5. $\begin{cases} 5x + 2y = 22 \\ 3x - 2y = 10 \end{cases}$

6. $\begin{cases} 5x + 3y = 1 \\ -2x + 5y = 12 \end{cases}$

7. $\begin{cases} 4x - 3y = -10 \\ 5x + 2y = 22 \end{cases}$

8. $\begin{cases} 2x + y = 0 \\ x - y = 1 \end{cases}$

9. $\begin{cases} -2x + 17y = 6 \\ 11x - 5y = -33 \end{cases}$

10. $\begin{cases} 5x + 2y = 50 \\ 4x - 3y = -52 \end{cases}$

11. $\begin{cases} 2x + \dfrac{1}{2}y = 2 \\ 6x - y = 1 \end{cases}$

12. $\begin{cases} y = \dfrac{-2}{5}x + 4 \\ x = \dfrac{1}{3}y - 7 \end{cases}$

13. $\begin{cases} 3x - 2y = 8 \\ -6x + 4y = 10 \end{cases}$

14. $\begin{cases} 2a + 3b = 10 \\ 3a + 2b = 10 \end{cases}$

15. $\begin{cases} 13x - 5y = -5 \\ x + y = 1 \end{cases}$

Solve these systems using elimination by substitution.

16. $\begin{cases} -7x + y = 2 \\ 2x - 8 = y \end{cases}$

17. $\begin{cases} x + 4y = 8 \\ y = \dfrac{-1}{4}x - 7 \end{cases}$

18. $\begin{cases} x - 3y = 2 \\ 4x - 10y = 10 \end{cases}$

19. $\begin{cases} x = 7 - \dfrac{1}{2}y \\ 8x - y = -4 \end{cases}$

20. $\begin{cases} y = \dfrac{-2}{7}x - 1 \\ 2x = -7(y + 1) \end{cases}$

- -

Solutions to Problems

1.

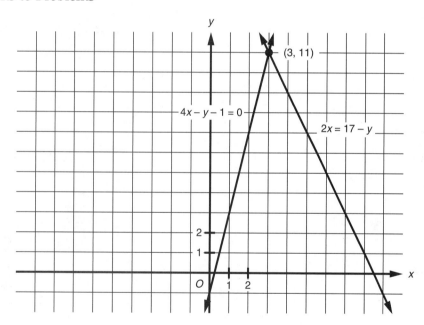

Answer: (3, 11) or $x = 3$ and $y = 11$

2. $4x - y - 1 = 0$
    $2x = 17 - y$

    $4x - y = 1$
    $2x + y = 17$
    $6x \quad\;\; = 18$
    $\quad\;\; x = 3$

If $x = 3$ and $2x = 17 - y$,
$$2(3) = 17 - y$$
$$6 = 17 - y$$
$$-11 = -y$$
$$11 = y$$

(3, 11)

3. $\quad x + 2y = 1$
$\quad 5x + 3y = 26$

$\quad -5(x + 2y = 1)$
$\quad 5x + 3y = 26$

$\quad -5x - 10y = -5$
$\quad 5x + 3y = 26$
$\quad\quad\quad -7y = 21$
$\quad\quad\quad\quad y = -3$

If $y = -3$ and $x + 2y = 1$,
$$x + 2(-3) = 1$$
$$x - 6 = 1$$
$$x = 7$$

(7, -3)

4. $\quad 3x - 2y = 5$
$\quad -4x + 3y = 1$

$\quad 4(3x - 2y = 5)$
$\quad 3(-4x + 3y = 1)$

$\quad 12x - 8y = 20$
$\quad -12x + 9y = 3$
$\quad\quad\quad\quad y = 23$

If $y = 23$ and $3x - 2y = 5$,
$$3x - 2(23) = 5$$
$$3x - 46 = 5$$
$$3x = 51$$
$$x = 17$$

(17, 23)

5. $\quad \dfrac{x}{3} + y = 3$

$\quad 4x + 3y = 0$

Clear fractions in first equation.

$$3\left(\frac{x}{3} + y = 3 \right)$$
$$x + 3y = 9$$

Multiply.

$\quad -4(x + 3y = 9)$
$\quad 4x + 3y = 0$

Add.

$\quad -4x - 12y = -36$
$\quad 4x + 3y = 0$
$\quad\quad\quad -9y = -36$
$\quad\quad\quad\quad y = 4$

If $y = 4$ and $4x + 3y = 0$,
$$4x + 3(4) = 0$$
$$4x + 12 = 0$$
$$4x = -12$$
$$x = -3$$

$(-3, 4)$

6. Create a table of values by solving each equation for y.

$$3x + 4y = 2$$
$$4y = -3x + 2$$
$$y = \frac{-3}{4}x + \frac{2}{4}$$

$$2y = 4 - \frac{3}{2}x$$
$$4y = 8 - 3x$$
$$y = 2 - \frac{3}{4}x$$

Select three convenient values for x. Suggestions for the first line are to let $x = -6, -2, 2,$ or 6. Suggestions for the second line are to let $x = -4, 0,$ and 4.

Solve for y-values by substitution. Plot the three points for each line and connect them with a straight line.

Answer:

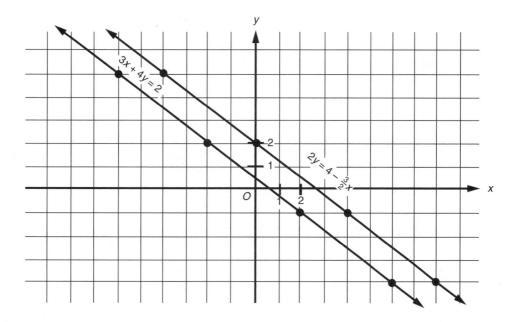

7. $5x - y = 3$
$2y = 10x - 6$

$$5x - y = 3$$
$$-10x + 2y = -6$$

$$2(5x - y = 3)$$
$$-10x + 2y = -6$$

$$10x - 2y = 6$$
$$-10x + 2y = -6$$
$$0 = 0$$

There are an infinite number of solutions that satisfy $y = 5x - 3$.

8. $4x + 3y = 9$

$$2y = 4 - \frac{8}{3}x$$

$$4x + 3y = 9$$

$$\frac{8}{3}x + 2y = 4$$

$$-2(4x + 3y = 9)$$

$$3\left(\frac{8}{3}x + 2y = 4\right)$$

$$-8x - 6y = -18$$
$$8x + 6y = 12$$
$$0 = -6$$

There is no solution to the system.

9. $3x + 5y = 14$

$$y = -2$$

Since $y = -2$:
$$3x + 5y = 14$$
$$3x + 5(-2) = 14$$
$$3x - 10 = 14$$
$$3x = 24$$
$$x = 8$$

$(8, -2)$

10. $y = 7x + 2$
$x - 3y = -6$

Since $y = 7x + 2$:

$$x \qquad - 3y = -6$$
$$x - 3(7x + 2) = -6$$
$$x - 21x - 6 = -6$$
$$-20x - 6 = -6$$
$$-20x = 0$$
$$x = 0$$

If $x = 0$ and $y = 7x + 2$,
$$y = 7(0) + 2$$
$$y = 0 + 2$$
$$y = 2$$

$(0, 2)$

UNIT 27

Solving Systems of Equations (Continued)

In Unit 26 you learned to solve systems of linear equations. When you have finished this unit, you will be able to solve systems of equations containing one linear equation and one quadratic equation.

First, recall how we define a solution:

Definition: The **solutions** to a system of equations are the ordered pairs of values of x and y that satisfy *all* the equations in the system.

In Unit 26, we saw that linear systems can be solved either graphically or algebraically. The same options apply to quadratic-linear systems.

GRAPHICAL SOLUTION OF QUADRATIC-LINEAR SYSTEM

Recall that the graph of a quadratic equation is a parabola and the graph of a linear equation is a line. A parabola and a line may intersect at two points, one point, or not at all. These intersection points, if any, are the solution(s) to the system.

To solve a quadratic-linear system graphically:

1. Graph the parabola and the line on the same set of axes. Label each graph.

2. Write the solution, the point or points where the two graphs intersect. (If the two graphs do not intersect, the system has no solution.)

EXAMPLE 1

Solve graphically and check: $\begin{cases} y = -x^2 + 2x + 4 \\ y = 4 - x \end{cases}$

Solution: 1.   We need to graph both equations.

a. $y = -x^2 + 2x + 4$ is a parabola. (See Unit 24 to review graphing quadratic equations.) The axis of symmetry is:

$$x = -\frac{b}{2a} = -\frac{2}{2(-1)} \rightarrow x = 1$$

x	$y = -x^2 + 2x + 4$	(x, y)
-2	$y = -(-2)^2 + 2(-2) + 4$ $= -4$	$(-2, -4)$
-1	$y = -(-1)^2 + 2(-1) + 4$ $= 1$	$(-1, 1)$
0	$y = -(0)^2 + 2(0) + 4$ $= 4$	$(0, 4)$
1	$y = -(1)^2 + 2(1) + 4$ $= 5$	$(1, 5)$
2	$y = -(2)^2 + 2(2) + 4$ $= 4$	$(2, 4)$
3	$y = -(3)^2 + 2(3) + 4$ $= 1$	$(3, 1)$
4	$y = -(4)^2 + 2(4) + 4$ $= -4$	$(4, -4)$

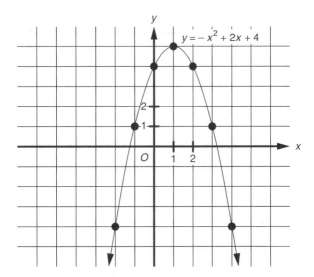

Remember to label the graphs when graphing more than one function on the same set of axes.

b. $y = 4 - x$ is a line (See Unit 23 to review graphing linear equations.) Rewrite the equation in $y = mx + b$ form:

$y = -x + 4$. The y-intercept is 4 and the slope is $-\dfrac{1}{1}$ (down one, right one).

In general, three (or even two) points are enough to graph a line, but when solving a system graphically by hand you may want to use more points to be sure the two graphs intersect at the correct points.

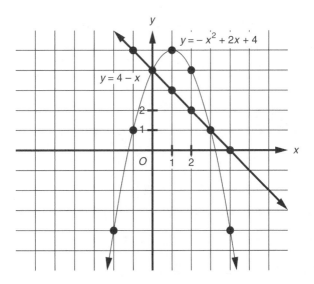

The solutions are the points where the two graphs intersect. In this problem, there are two solutions: (0, 4) and (3, 1).

Since this problem has two solutions, each of which must work in two equations, a complete check of the problem involves four individual checks. Consider organizing the checks in a table.

	$y = -x^2 + 2x + 4$	$y = 4 - x$
(0, 4)	$4 = -(0)^2 + 2(0) + 4$ $4 = 4$ ✓	$4 = 4 - (0)$ $4 = 4$ ✓
(3, 1)	$1 = -(3)^2 + 2(3) + 4$ $1 = 1$ ✓	$1 = 4 - (3)$ $1 = 1$ ✓

These solutions both check in both equations.

Answer: (0, 4) or (3, 1)

The example above shows that a quadratic-linear system in two variables may have exactly two solutions. You should be able to convince yourself that a line can never intersect a parabola in more than two points, so two is the maximum number of solutions for such a system. Problem 1 for you shows that a quadratic-linear system can have just one solution. It is also possible for there to be no solution; this happens when the line and parabola do not intersect.

Problem 1

Solve graphically and check: $\begin{cases} y = \dfrac{1}{2}x^2 + 2x + 3 \\ y = -2x - 5 \end{cases}$

Hint: Use even x-values for the parabola. Five points will be enough.

Solution:

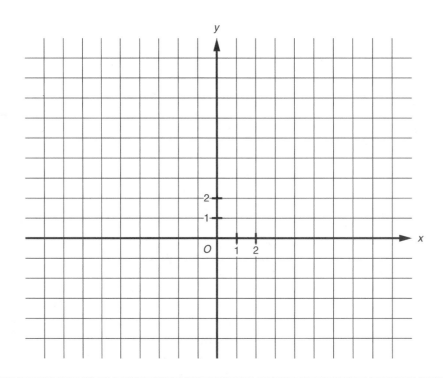

ALGEBRAIC SOLUTION OF QUADRATIC-LINEAR SYSTEM

As in Unit 26, our strategy for solving a system of equations is to eliminate one variable. For a system of two linear equations, either elimination by addition or elimination by substitution can be used. For a quadratic-linear system, the preferred method is substitution.

1. Use the quadratic equation to **substitute** into the linear equation. If necessary, write the quadratic equation in standard form first, $y = ax^2 + bx + c$.

2. **Solve** the equation from step 1 either by factoring or by using the quadratic formula.

3. **Substitute** the answer (5) back into one of the original equations, and solve for the second variable.

   The examples that follow illustrate the use of these steps in solving a system of equations containing one quadratic equation.

EXAMPLE 2

Solve: $\begin{cases} y = 2x + 3 \\ y = x^2 \end{cases}$.

- -

Solution: Use $y = x^2$ to substitute into the linear equation.

$$y = 2x + 3$$

$$\downarrow$$

$$x^2 = 2x + 3 \qquad \text{Replace } y \text{ with } x^2.$$

$$x^2 - 2x - 3 = 0 \qquad \text{Solve the new equation.}$$

$$(x + 1)(x - 3) = 0$$

$$x + 1 = 0 \qquad \text{or} \qquad x - 3 = 0$$

$$x = -1 \qquad \text{or} \qquad x = 3$$

Substitute back into an original equation.

If $x = -1$	If $x = 3$
and $y = x^2$,	and $y = x^2$,
$y = (-1)^2$,	$y = (3)^2$,
$y = 1.$	$y = 9.$

Answer: There are two solutions to the system:

$$x = -1 \quad \text{and} \quad y = 1 \quad (-1, 1)$$

$$\text{and}$$

$$x = 3 \quad \text{and} \quad y = 9 \quad (3, 9)$$

- -

EXAMPLE 3

Solve: $\begin{cases} y = -x^2 + 2x + 3 \\ y = 3 - x \end{cases}$.

Solution: Use $y = -x^2 + 2x + 3$ to substitute into the equation.

$$y = 3 - x$$

$-x^2 + 2x + 3 = 3 - x$ Solve the new equation.

$$-x^2 + 3x = 0$$
$$-1(-x^2 + 3x = 0)$$
$$x^2 - 3x = 0$$
$$x(x - 3) = 0$$
$$x = 0 \quad \text{or} \quad x - 3 = 0$$
$$x = 3$$

Substitute back into an original equation.

If $x = 0$	If $x = 3$
and $y = 3 - x$,	and $y = 3 - x$,
$= 3 - 0$,	$= 3 - 3$,
$= 3$.	$= 0$.

Answer: There are two solutions to the system:

$$x = 0 \quad \text{and} \quad y = 3 \qquad (0, 3)$$
$$\text{and}$$
$$x = 3 \quad \text{and} \quad y = 0 \qquad (3, 0).$$

Try solving some yourself.

Problem 2

Solve algebraically: $\begin{cases} y = x^2 + x + 1 \\ y = -x \end{cases}$.

Solution:

Problem 3

Solve algebraically: $\begin{cases} y = -x^2 + 2x + 8 \\ x - y = -6 \end{cases}$

EXAMPLE 4

Solve: $\begin{cases} y = x^2 + 5x + 4 \\ 2x - 3y = 15 \end{cases}$.

Solution: Use $y = x^2 + 5x + 4$ to substitute into the equation.

$$2x - \qquad 3y \qquad = 15$$

$$2x - 3(\overbrace{x^2 + 5x + 4}) = 15$$

$$2x - 3x^2 - 15x - 12 = 15$$

$$-3x^2 - 13x - 27 = 0$$

$$3x^2 + 13x + 27 = 0$$

Not readily factorable; use quadratic formula with $a = 3$, $b = 13$, and $c = 27$.

$$x = \frac{-b \pm \sqrt{b^2 - 4ac}}{2a}$$

$$= \frac{-(13) \pm \sqrt{(13)^2 - 4(3)(27)}}{2(3)}$$

$$= \frac{-13 \pm \sqrt{169 - 324}}{6}$$

$$= \frac{-13 + \sqrt{-155}}{6}$$

No solution.

Answer: There is no solution to the system.

Recall what was stated in Unit 26 regarding no solution to a system: The equations are said to be **inconsistent** when there is no solution to the system.

Graphically, no solution to a system of equations means that there is no point of intersection for the two graphs.

Verify, by doing Problem 4, that the system in Example 4 represents a line and a parabola that do not intersect.

Problem 4

Graph: $\begin{cases} y = x^2 + 5x + 4 \\ 2x - 3y = 15 \end{cases}$.

Solution:

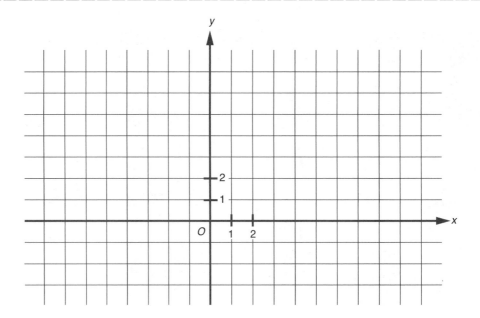

Here are a few more systems for you to solve. For additional practice, you might consider graphing them as well.

Problem 5

Solve: $\begin{cases} y = -x^2 + 4x + 5 \\ x - y = -5 \end{cases}$.

Solution:

Problem 6

Solve: $\begin{cases} y = x^2 + 2 \\ 3x + 4y = -8 \end{cases}$.

Solution:

ALGEBRA AND THE CALCULATOR (Optional)

Finding Points of Intersection

We will revisit Problem 2, but this time we will find the solution using a graphing calculator.

EXAMPLE 5

Given: $\begin{cases} y = x^2 + x + 1 \\ y = -x \end{cases}$.

Graph and find the point of intersection.

Solution: Use the standard viewing window $\boxed{\text{ZOOM}}\ \boxed{6}$.

Key in *both* equations as follows:

$\boxed{\text{Y=}}\ \boxed{\text{CLEAR}}\ \boxed{\text{X,T,}\theta\text{,n}}\ \boxed{x^2}\ \boxed{+}\ \boxed{\text{X,T,}\theta\text{,n}}\ \boxed{+}\ \boxed{1}\ \boxed{\text{ENTER}}$.

$\boxed{\text{Y=}}\ \boxed{\text{CLEAR}}\ \boxed{(-)}\ \boxed{\text{X,T,}\theta\text{,n}}\ \boxed{\text{ENTER}}$.

and view the graph. The graph should match the one shown in Problem 3.

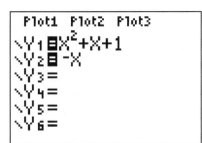
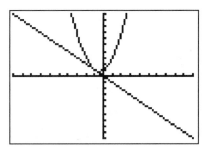

To find the point of intersection, press $\boxed{\text{2nd}}\ \boxed{\text{CALC}}\ \boxed{5}$.

At the <u>First curve?</u> prompt, press $\boxed{\text{ENTER}}$.

At the <u>Second curve?</u> prompt, press $\boxed{\text{ENTER}}$.

At the <u>Guess?</u> prompt, move the cursor to where the intersection appears to be, and press ENTER.

Your screen may look like one of the screens below, or it may be slightly different. The intersection method on the calculator can have slight errors, usually at the fourth decimal place or smaller. If you round your answer to the nearest tenth or hundredth, all the solutions will be equivalent. These intersection points both round to $x = -1$ and $y = 1$.

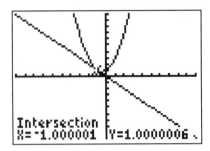 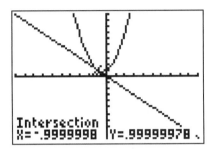

Answer: Intersection $X = -1$ $Y = 1$ appears at the bottom of the screen.

_____

You should now be able to solve any system of equations containing one linear equation and one quadratic equation.

To solve a quadratic-linear system graphically:

1. Graph the parabola and the line on the same set of axes. Label each graph.
2. Write the solution, the point or points where the two graphs intersect. (If the two graphs do not intersect, the system has no solution.)

Remember that elimination by substitution is the algebraic method used to solve such systems; briefly stated, the three basic steps involved are:

1. **Substitute into** the linear equation.
2. **Solve.**
3. **Substitute back.**

Before beginning the next unit you should solve the following systems. Optional—verify your answers by reworking the problems on a graphing calculator before checking your answers with the back of the book.

====================================

EXERCISES

Solve these systems using elimination by substitution.

1. $\begin{cases} y = 3x^2 - 7x + 11 \\ x = 2 \end{cases}$

2. $\begin{cases} y = 4x^2 \\ -4x + y = 0 \end{cases}$

3. $\begin{cases} y = x^2 + 5x \\ y = -2x + 44 \end{cases}$

4. $\begin{cases} y = x^2 + 5x - 21 \\ y = x \end{cases}$

5. $\begin{cases} y = x^2 + 2x + 5 \\ -4x + 2y = -7 \end{cases}$

6. $\begin{cases} y = 2x^2 - 4x \\ y = 8x - 18 \end{cases}$

7. $\begin{cases} y = 3x^2 + 8x - 2 \\ y = 1 \end{cases}$

8. $\begin{cases} y = x^2 + 13 \\ y = -10x - 12 \end{cases}$

Solve these systems graphically.

9. $\begin{cases} y = -x^2 + 2x + 8 \\ x - y = -6 \end{cases}$

10. $\begin{cases} y = x^2 - 4 \\ y + 2x = -1 \end{cases}$

Solutions to Problems

1. $y = \dfrac{1}{2}x^2 + 2x + 3$

$y = -2x - 5$

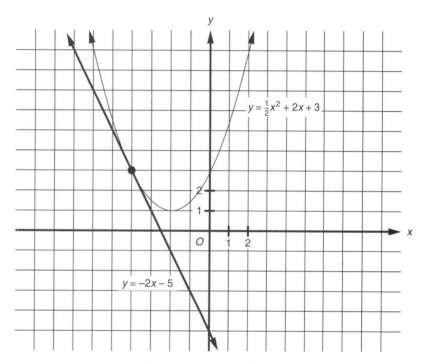

Answer: $(-4, 3)$

2. $\qquad\qquad y = x^2 + x + 1$

$\qquad\qquad\quad y = -x$

$\qquad\quad x^2 + x + 1 = -x$

$\qquad\quad x^2 + 2x + 1 = 0$

$\qquad (x + 1)(x + 1) = 0$

$x + 1 = 0 \qquad$ or $\qquad x + 1 = 0$

$\quad x = -1 \qquad\qquad\qquad x = -1$

$\quad x = -1 \qquad\qquad\qquad x = -1$

$\quad y = -x \qquad\qquad\qquad y = -x$

$\quad y = -(-1) \qquad\qquad\quad y = -(-1)$

$\quad y = 1 \qquad\qquad\qquad\quad y = 1$

There is one solution to the system:

$\qquad\qquad x = -1$ and $y = 1 \qquad (-1, 1)$

3. $y = -x^2 + 2x + 8$
                             $x - y = -6$
   $x - (-x^2 + 2x + 8) = -6$
      $x + x^2 - 2x - 8 = -6$
            $x^2 - x - 2 = 0$
         $(x - 2)(x + 1) = 0$

   $x - 2 = 0$ or $x + 1 = 0$
     $x = 2$ $x = -1$

     $x = 2$ $x = -1$
   $x - y = -6$ $x - y = -6$
   $2 - y = -6$ $-1 - y = -6$
    $-y = -8$ $-y = -5$
     $y = 8$ $y = 5$

   There are two solutions to the system:

                    $x = 2$ and $y = 8$ (2, 8)
                    and
                    $x = -1$ and $y = 5$ (–1, 5)

4. $y = x^2 + 5x + 4$; vertex $\left(\dfrac{-5}{2}, \dfrac{-9}{4} \right)$;

    $2x - 3y = 15 \longrightarrow y = \dfrac{2}{3}x - 5$

    Answer:

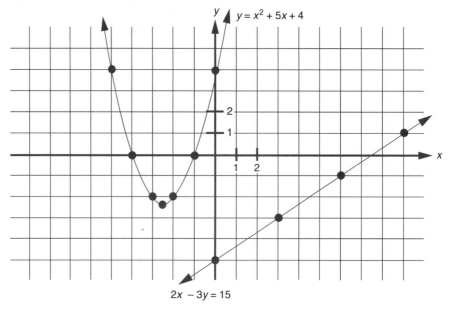

5.
$$y = -x^2 + 4x + 5$$
$$x - y = -5$$
$$x - (-x^2 + 4x + 5) = -5$$
$$x + x^2 - 4x - 5 = -5$$
$$x^2 - 3x - 5 = -5$$
$$x^2 - 3x = 0$$
$$x(x - 3) = 0$$

$$x = 0 \qquad \text{or} \qquad x - 3 = 0$$
$$x = 3$$

$x = 0$	$x = 3$
$x - y = -5$	$x - y = -5$
$0 - y = -5$	$3 - y = -5$
$y = 5$	$-y = -8$
	$y = 8$

There are two solutions to the system:
$$x = 0 \quad \text{and} \quad y = 5 \qquad (0, 5)$$
and
$$x = 3 \quad \text{and} \quad y = 8 \qquad (3, 8)$$

6.
$$y = x^2 + 2$$
$$3x + 4y = -8$$
$$3x + 4(x^2 + 2) = -8$$
$$3x + 4x^2 + 8 = -8$$
$$4x^2 + 3x + 16 = 0$$

$$x = \frac{-b \pm \sqrt{b^2 - 4ac}}{2a}$$

$$x = \frac{-(3) \pm \sqrt{(3)^2 - 4(4)(16)}}{2(4)}$$

$$x = \frac{-3 \pm \sqrt{9 - 256}}{8}$$

$$x = \frac{-3 \pm \sqrt{-247}}{8}$$

There is no solution to this system.

UNIT 28

Solving Inequalities—First-Degree

The purpose of this unit is to provide you with an understanding of inequalities. When you have finished the unit, you will be able to solve all first-degree inequalities.

INEQUALITY SIGNS

There are two inequality signs:

> < read "less than"
>
> > read "greater than"

Sometimes an inequality sign and an equal sign are combined:

> ≤ read "less than or equal to"
>
> ≥ read "greater than or equal to"

For our purposes in this book we will use a "commonsense" definition of <, based on your familiarity with the number line. Remember the number line? It looks like this:

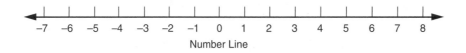

Number Line

> Definition: $a < b$ if a is to the left of b on the number line.

Here are some examples:

$2 <\ 3$ because 2 is to the left of 3 on the number line.

$-2 <\ 5$ because -2 is to the left of 5 on the number line.

$-5 < -2$ because -5 is to the left of -2.

$-7 < -6$ because -7 is to the left of -6.

The sign > can be defined in a similar manner.

> Definition: $a > b$ if a is to the right of b on the number line.

For instance:

$3 >\ 2$ because 3 is to the right of 2.

$0 > -1$ because 0 is to the right of -1.

$-4 > -5$ because -4 is to the right of -5.

Notice that the inequality sign always opens to the larger number.

$b > a$ is equivalent to $a < b$.

Observe that $5 > 2$ has the same meaning as $2 < 5$. Both statements indicate that 5 is the larger number and 2 is the smaller number. Thus:

$3 > -1$ can be rewritten as $-1 <\ 3$.

$-8 < -2$ can be rewritten as $-2 > -8$.

$3 >\ x$ can be rewritten as $\ x <\ 3$.

$0 <\ y$ can be rewritten as $\ y >\ 0$.

FIRST-DEGREE INEQUALITIES

Remember that in any first-degree equation in one variable:

There is only one variable.

The variable is involved in only the four fundamental arithmetic operations.

The variable is never multiplied by itself.

The variable is never in a denominator.

A **first-degree inequality** in one variable has the same characteristics as a first-degree equation except that in place of the equal sign there is an inequality sign.

Here are some examples of first-degree inequalities:

$$2x + 5 < 7$$
$$3(x - 1) + x \geq 2 - x$$
$$1 - x \leq 15(3 + 2x) - x$$

To solve a first-degree inequality, find the values of x that satisfy the inequality. The basic strategy is the same as that used to solve first-degree equations—get all terms involving x on one side of the inequality sign, and get all other terms on the other side.

To accomplish this, we use two rules:

Rule 1: **A term may be transposed from one side of the inequality to the other by changing its sign as it crosses the inequality sign.**

EXAMPLE 1　If $x + 5 < 7$,

then $x < 7 - 5$

Note sign change.

and $x < 2$.

EXAMPLE 2　If $1 - x > -6$,

then $1 + 6 > x$

Note sign changes.

and $7 > x$

or $x < 7$.

Note: Writing the inequality with x on the left makes it easier to read.

Rule 2: **Both sides of an inequality may be multiplied or divided by the same nonzero number provided that:**
a.　if the number is positive, the direction of the inequality remains the same;
b.　if the number is negative, the direction of the inequality is reversed.

EXAMPLE 3　　　$6 < 15$,

so $\dfrac{6}{3} < \dfrac{15}{3}$

and $2 < 5$.

EXAMPLE 4　　　$\dfrac{1}{4} < 12$,

so $4\left(\dfrac{1}{4}\right) < 4(12)$

and $1 < 48$.

EXAMPLE 5　　　$15 > 10$,

so $\dfrac{15}{-5} < \dfrac{10}{-5}$

and $-3 < -2$.

Note: For the inequality to be true, the direction of the **inequality must be reversed** when the inequality is divided by the same negative number.

EXAMPLE 6 If $-2x \le 8$,

then $\dfrac{-2x}{-2} \ge \dfrac{8}{-2}$

and $x \ge -4.$

Note again: For the inequality to be true, the direction of the **inequality must be reversed** when the inequality is divided by the same negative number.

EXAMPLE 7 If $-x > -2$,

then $(-1)(-x) < (-1)(-2)$

and $-x < 2.$

Note: In this example, the inequality is multiplied by the same negative number, and for the inequality to be true, the direction of the **inequality must be reversed**.

Now let's get on with the business of solving first-degree inequalities. The same four steps used to solve first-degree equations will again be used to solve inequalities. However, we must remember to change the direction of the inequality if we multiply or divide by the same negative number.

Recall the four steps:

> 1. **Simplify** by removing parentheses,
>   clearing fractions,
>   collecting like terms.
> 2. **Transpose.**
> 3. **Simplify.**
> 4. **Divide by the coefficient of the variable.**

Now, we must remember:

> **Reverse the direction** of an inequality symbol whenever an inequality is multiplied or divided by the same negative number.

Here are some examples that illustrate the use of these steps in solving first-degree inequalities. Read through each example and be sure you understand what has happened at each step.

EXAMPLE 8

Solve: $4x - 7 > 6x + 5$.

Solution:

$$4x - 7 > 6x + 5$$
$$4x - 6x > 5 + 7$$
$$-2x > 12$$
$$\dfrac{-2x}{-2} < \dfrac{12}{-2} \quad \text{Reverse direction.}$$
$$x < -6$$

Answer: $x < -6$

To say that $x < -6$ is the answer to the inequality means that all numbers less than -6 satisfy the inequality. For example, if we substitute $x = -7$ into the inequality, we obtain

$$4(-7) - 7 > 6(-7) + 5$$
$$-28 - 7 > -42 + 5$$
$$-35 > -37$$

and, rewriting, $\qquad -37 < -35$

which is true, so we say that $x = -7$ satisfies the inequality.

Graphically, $x < -6$ can be represented on the number line with an open circle at -6 and a heavy line to the left. The open circle indicates that -6 is not part of the answer. The heavy line indicates that all numbers to the left of -6 are part of the answer, including such numbers as -6.1 and -6.25.

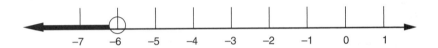

EXAMPLE 9

Solve: $5(x + 3) \geq 31 + x$.

Solution:
$$5(x + 3) \geq 31 + x$$
$$5x + 15 \geq 31 + x$$
$$5x - x \geq 31 - 15$$
$$4x \geq 16$$
$$x \geq 4$$

Answer: $x \geq 4$

To say that $x \geq 4$ is the answer to the inequality means that all numbers greater than *or equal to* 4 will satisfy the inequality.

If the answer were graphed on a number line, there would be a closed circle at 4 to indicate that $x = 4$ is part of the solution and a heavy line to the right.

Now try to solve the inequalities in Problems 1 and 2. Remember to **reverse the direction of the inequality symbol** whenever the inequality is multiplied or divided by the same negative number.

Problem 1

Solve: $3x - x - 7 \geq (x - 2) + 4$ and sketch the solution on a number line.

Solution:

Problem 2

Solve: $x - 4 - 2(6 - x) > 2(3x - 5)$ and sketch the solution on the number line.

Solution:

Now, in contrast to our earlier examples, consider the following inequality.

EXAMPLE 10

Solve: $2x + 1 + x < 3(x + 2)$ and sketch the solution on a number line.

Solution: $2x + 1 + x < 3(x + 2)$

$$1 + 3x < 3x + 6$$

$$3x - 3x < 6 - 1$$

$$0 < 5$$

Note: $0 < 5$ is always true regardless of the value of x.

Answer: The solution is the entire set of real numbers.

The inequality in Example 10 is *always* true, regardless of the value of x, because 0 is always less than 5. Its solution is the entire set of real numbers. Remember, the real numbers are all the counting numbers, zero, whole numbers, positive and negative integers, fractions, rational numbers, and irrational numbers. In other words, x can equal any number.

For example, if $x = 0$, then $2(0) + 1 + (0) < 3((0) + 2)$

and $1 < 6$.

Or:
$$\text{if } x = 5, \text{ then } 2(5) + 1 + (5) < 3((5) + 2)$$
$$\text{and} \quad 10 + 1 + 5 < 3(7)$$
$$16 < 21.$$

You should now be able to solve any first-degree inequality.

The same four basic steps—simplify, transpose, simplify, and divide—are used to solve first-degree inequalities. In addition, you must remember to reverse the direction of the inequality symbol whenever you multiply or divide an inequality by the same negative number.

Now try to solve the problems in the exercises.

EXERCISES

Solve, and graph the answer for each problem on a number line:

1. $10 + 2x > 12$

2. $4 - (12 - 3x) \leq -5$

3. $5x < 22 - (2x + 1)$

4. $4x + (3x - 7) > 2x - (28 - 2x)$

5. $5 - 3x \leq 23$

6. $3x + 4(x - 2) \geq x - 5 + 3(2x - 1)$

7. $3x - 2(5x + 2) > 1 - 5(x - 1) + x$

8. $3x - 2(x - 5) < 3(x - 1) - 2x - 11$

9. $3x + 4(x - 2) + 7 > x - 5 + 3(2x - 1)$

10. $5x - 2(3x - 4) > 4[2x - 3(1 - 3x)]$

Solutions to Problems

1. $3x - x - 7 \geq (x - 2) + 4$
    $2x - 7 \geq x + 2$
    $x - 7 \geq 2$
    $x \geq 9$

9

2. $x - 4 - 2(6 - x) > 2(3x - 5)$
    $x - 4 - 12 + 2x > 6x - 10$
    $3x - 16 > 6x - 10$
    $-3x - 16 > -10$
    $-3x > 10$
    $x < -2$

−2

UNIT 29

Solving Quadratic Inequalities

In this unit we will continue our discussion of inequalities. When you have completed the unit, you will be able to solve quadratic inequalities in one variable.

WRITING DOUBLE INEQUALITIES

A **double inequality** has two inequality symbols of the same sense combined in one statement.

For example, $2 < x < 5$ is a double inequality. In words, $2 < x < 5$ means that x represents all numbers greater than 2 but less than 5. In other words, x lies between 2 and 5 on the number line.

Graphically, $2 < x < 5$ is represented as follows:

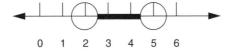

Observe that $5 > x > 2$ has the same meaning as $2 < x < 5$. Both statements indicate that 5 is the larger number, 2 is the smaller number, and x lies between them.

We prefer to write double inequalities using the $<$ symbol, because it is clearer to visualize the relationship of the numbers and the symbols on the number line. For example:

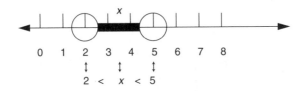

What does $-1 \le x \le 4$ mean? To say that $-1 \le x \le 4$ indicates that x represents all numbers between -1 and 4, including the numbers -1 and 4. Graphically it would look like this:

SECOND-DEGREE OR QUADRATIC INEQUALITIES

Recall from Unit 21:

Definition: $ax^2 + bx + c = 0$, with a, b, and c being real numbers, $a \ne 0$, is called a **second-degree equation** or **quadratic equation**.

A second-degree or quadratic inequality has the same characteristics as a second-degree equation, except that in place of the equal sign there is an inequality symbol. The standard form of the inequality has **all** terms on the left side of the inequality symbol with **only** 0 on the right.

Examples of second-degree or quadratic inequalities in standard form are:

$$x^2 - 3x + 5 < 0$$
$$7x^2 + 2x - 1 > 0$$
$$x^2 \ge 0$$
$$4x^2 + 7 \le 0$$

whereas

$$x^3 - x^2 > 0 \text{ is a third-degree inequality}$$
$$x^2 + y < 0 \text{ has two variables}$$
$$3x - 7 \le x \text{ is a first-degree inequality}$$

To solve a second-degree or quadratic inequality is to find the values of x that satisfy the inequality. The basic technique developed here will rely heavily on our knowledge of graphing quadratic or second-degree equations.

Recall some of the basic facts about the graph of a quadratic equation in standard form: $y = ax^2 + bx + c$.

The graph is a smooth, $\smile$-shaped curve called a parabola.

If a is positive, the curve opens up. If a is negative, the curve opens down.

There are at most two x-intercepts, which are found by solving $0 = ax^2 + bx + c$.

The procedure we will use to solve quadratic inequalities involves four major steps.

1. If necessary, first rewrite the inequality in standard form; that is, put all the terms on the left side with only 0 remaining on the right side of the inequality symbol.

2. Let y = the algebraic expression on the left side of the inequality symbol, resulting in a quadratic **equation**.

3. Find the x-intercepts by solving $0 = ax^2 + bx + c$ and do a rough sketch of the quadratic equation showing only the x-intercept(s). There is no need to locate the vertex or any extra points.

4. By inspection of the graph, determine the answer to the inequality.

The examples that follow illustrate this procedure.

EXAMPLE 1

Solve: $x^2 - 2x - 3 < 0$.

Solution: Let $y = x^2 - 2x - 3$ and graph.

This is a quadratic equation; graph is a parabola.

Since $a = 1$ is positive, curve opens up.

Find the x-intercepts, at most two, by solving:

$$0 = x^2 - 2x - 3$$
$$0 = (x + 1)(x - 3)$$
$$x = -1 \quad \text{or} \quad x = 3$$

The x-intercepts are -1 and 3.

Do a rough sketch of the parabola, showing only the x-intercepts. There is no need to locate the vertex.

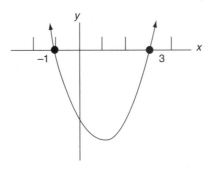

By inspection of the graph, answer this question:

For what values of x is $x^2 - 2x - 3 < 0$?

Or, in other words:

For what values of x is $y < 0$?

Or, in still other words:

For what values of x is the curve below the x-axis?

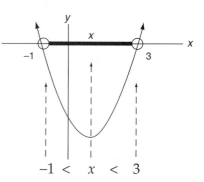

The curve is below the x-axis when x is any number between -1 and 3.

Answer: $-1 < x < 3$

$$-1 < \quad x \quad < 3$$

To say that $-1 < x < 3$ is the answer to the inequality means that all numbers between -1 and 3 satisfy the inequality. For example, if we substitute $x = 1$ into the inequality, we obtain

$$(1)^2 - 2(1) - 3 < 0$$
$$1 - 2 - 3 < 0$$
$$-4 < 0$$

which is true, so we say that $x = 1$ satisfies the inequality.

EXAMPLE 2

Solve: $x^2 + 5x + 6 < 0$.

Solution: Let $y = x^2 + 5x + 6$ and graph.

This is a quadratic equation; graph is a parabola.

Since $a = 1$ is positive, curve opens up.

Find the x-intercepts, at most two, by solving:

$$0 = x^2 + 5x + 6$$
$$0 = (x + 2)(x + 3)$$
$$x = -2 \quad \text{or} \quad x = -3$$

The x-intercepts are -2 and -3.

Do a rough sketch of the parabola, showing only the x-intercepts. There is no need to locate the vertex.

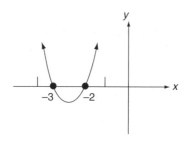

By inspection of the graph, answer this question:

$$\text{For what values of } x \text{ is } x^2 + 5x + 6 < 0?$$

Or:

$$\text{For what values of } x \text{ is } y < 0?$$

Or:

$$\text{For what values of } x \text{ is the curve below the } x\text{-axis?}$$

The curve is below the x-axis when x is any number between -3 and -2.

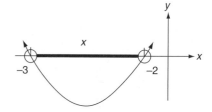

Answer: $-3 < x < -2$

$$-3 \quad < \quad x \quad < \quad -2$$

Try solving this problem.

Problem 1

Solve: $x^2 - 3x - 10 < 0$.

Solution:

Will do another example, then you try two problems.

EXAMPLE 3

Solve: $-x^2 + 4 < 0$.

Solution: Let $y = -x^2 + 4$ and graph.

Since $a = -1$ is negative, the parabola opens down.

Find the x-intercepts, at most two, by solving:

$$0 = -x^2 + 4$$
$$x^2 = 4$$
$$x = -2 \qquad \text{or} \qquad x = 2$$

The x-intercepts are -2 and 2.

Do a rough sketch of the parabola, showing only x-intercepts.

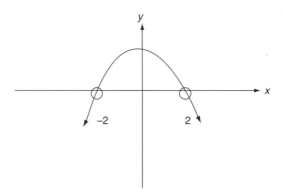

By inspection of the graph, answer this question:

For what values of x is the curve below the x-axis?

In this example, the parabola is opening down, and the two ends of the curve are below the x-axis. Thus the curve is below the x-axis when x is less than -2 or when x is greater than 2.

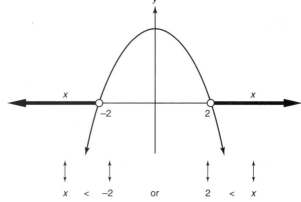

Answer: $x < -2$ or $x > 2$

Note: The answer cannot be written as a double inequality because x does *not* lie between two numbers.

Problem 2

Solve: $-x^2 + 4x + 5 < 0$.

Solution:

Problem 3

Solve: $x^2 + 1 < 0$.

Solution:

Thus far, all our examples have been inequalities with the < symbol. When an inequality has a > symbol, the procedure for solving it is similar except for the final question.

To solve an inequality, written in standard form, involving >, the final question becomes:

For what values of x is $y > 0$?

Or, in other words:

For what values of x is the curve **above** the x-axis?

EXAMPLE 4

Solve: $-x^2 + x + 12 > 0$.

Solution: Let $y = -x^2 + x + 12$ and graph.

This is a quadratic equation; graph is a parabola.

Since $a = -1$ is negative, curve opens down.

The x-intercepts:

$$0 = -x^2 + x + 12$$
$$0 = x^2 - x - 12$$
$$0 = (x - 4)(x + 3)$$
$$x = 4 \quad \text{or} \quad x = -3$$

The x-intercepts are -3 and 4.

A rough sketch of the parabola is:

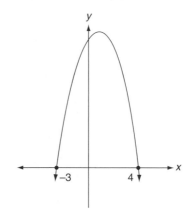

By inspection of the graph, answer this question:

For what values of x is $-x^2 + x + 12 > 0$?

Or:

For what values of x is $y > 0$?

Or:

For what values of x is the curve **above** the x-axis?

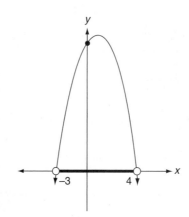

Answer: $-3 < x < 4$

Here is an inequality with > for you to solve.

Problem 4

Solve: $-x^2 + 7x - 6 > 0$.

_____

Solution:

_ _

To generalize the situation:

Given $ax^2 + bx + c < 0$ to solve, the final question to be answered is:

For what values of x is the curve **below** the x-axis?

Given $ax^2 + bx + c > 0$ to solve, the final question to be answered is:

For what values of x is the curve **above** the x-axis?

Here is one more example in detail with explanation.

EXAMPLE 5

Solve: $3x^2 \geq -4x - 1$.

_____

Solution: Write in standard form: $3x^2 + 4x + 1 \geq 0$.

Let $y = 3x^2 + 4x + 1$ and graph.

This is a quadratic equation; graph is a parabola.

Since $a = 3$ is positive, curve opens up.

The x-intercepts:

$$0 = 3x^2 + 4x + 1$$
$$0 = (3x + 1)(x + 1)$$
$$x = -\frac{1}{3} \quad \text{or} \quad x = -1$$

The x-intercepts are -1 and $-\dfrac{1}{3}$.

A rough sketch of the parabola is:

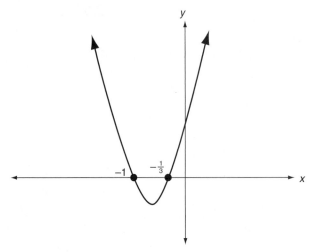

By inspection of the graph, answer this question:

For what values of x is the curve **above** or **on** the x-axis?

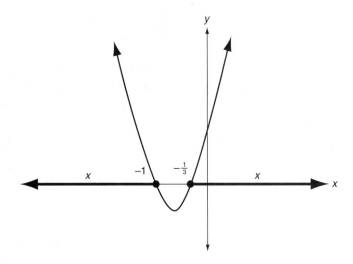

Answer: $x \leq -1$ or $x \geq \dfrac{-1}{3}$

Note: The x-intercepts are part of the answer. Why?

Convince yourself, by substitution, that both $x = -1$ and $x = -\dfrac{1}{3}$ satisfy the inequality.

– –

You should now be able to identify and solve second-degree or quadratic inequalities in one variable. Remember our four-step procedure, which, simply stated, is:

1. Write in standard form.
2. Let $y =$ the algebraic expression on the left.
3. Draw a rough sketch showing only the intercept(s).
4. By inspection, determine the answer.

If the inequality involves <, the final question to be answered is:

For what values of x is the curve **below** the x-axis?

If the inequality involves >, the final question to be answered is:

For what values of x is the curve **above** the x-axis?

Also remember, if the inequality is combined with an = sign, the x-intercepts are part of the answer.

Before going on to the next unit, try solving the following exercises.

EXERCISES

Solve for x by sketching the graph of each quadratic equation. Or, optionally, view the graph of each quadratic equation on a graphing calculator and determine its x-intercept(s). Then by inspection of the graph, determine the solution to the inequality.

1. $x^2 + 4x - 21 < 0$

2. $x^2 + 2x - 8 < 0$

3. $x^2 - x - 12 > 0$

4. $-x^2 - 4x - 3 < 0$

5. $-x^2 - x + 2 > 0$

6. $-4x^2 \geq 0$

7. $x^2 - 4x + 4 \leq 0$

8. $4 + 2x^2 \leq 0$

9. $9x^2 - 9x \leq 0$

10. $25 - x^2 \geq 0$

11. $x^2 + 1 > 0$

12. $x^2 + 6x + 9 > 0$

13. $-x^2 + 2x + 15 \leq 0$

14. $2x^2 + 1 \leq 0$

15. $-x^2 + 10x \leq 0$

16. $2x^2 + 3x - 5 \leq 0$

17. $-6x^2 - x + 2 \geq 0$

Solutions to Problems

1. Let $y = x^2 - 3x - 10$ and graph. Since $a = 1$, the parabola opens up. The x-intercepts are -2 and 5.

    $x^2 - 3x - 10 = 0$

    $(x - 5)(x + 2) = 0$

    $x = 5$ or $x = -2$

    Make a rough sketch.

    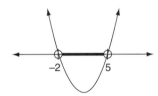

    The curve is below the x-axis ($y < 0$) for $-2 < x < 5$.

2. Let $y = -x^2 + 4x + 5$ and graph. Since $a = -1$, the parabola opens down. The x-intercepts are -1 and 5.

    $-x^2 + 4x + 5 = 0$

    $x^2 - 4x - 5 = 0$

    $(x - 5)(x + 1) = 0$

    $x = 5$ or $x = -1$

    Make a rough sketch.

    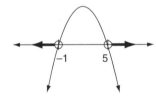

    The curve is below the x-axis ($y < 0$) for $x < -1$ or $x > 5$.

3. Let $y = x^2 + 1$ and graph. Since $a = 1$, the parabola opens up. There are no x-intercepts. The entire graph is above the x-axis.

    $x^2 + 1 = 0$

    $x^2 = -1$

    Make a rough sketch.

    The curve is never below the x-axis ($y < 0$). There is no solution.

4. Let $y = -x^2 + 7x - 6$ and graph. Since $a = -1$, the parabola opens down. The x-intercepts are 1 and 6.

$$-x^2 + 7x - 6 = 0$$

$$x^2 - 7x + 6 = 0$$

$$(x - 6)(x - 1) = 0$$

$$x = 6 \text{ or } x = 1$$

Make a rough sketch.

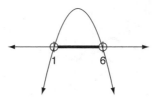

The curve is above the x-axis ($y > 0$) for $1 < x < 6$.

_____

UNIT 30

Absolute Value Equations, Inequalities, and Graphs

In this unit you will learn to work with absolute value equations and inequalities. When you are done with this unit, you should be able to solve simple absolute value equations and inequalities in one variable and to graph simple absolute value equations in two variables.

SOLVING ABSOLUTE VALUE EQUATIONS

An absolute value equation in one variable is an equation where the variable appears inside absolute value symbols. Here are some examples:

$$|x| = 5$$
$$|2x - 3| = 7$$
$$3|x + 4| = x - 2$$

While the method introduced below for solving absolute value equations can be applied to more complicated equations, in this book we stick to equations where the highest exponent is 1. We will not solve equations like.

$$|x^2 - 2x - 4| = 4.$$

In Unit 1, we said that the absolute value of a signed number is the number that remains when the sign is removed. Before solving absolute value equations, it is helpful to have a more precise definition of absolute value. For any number x,

$$|x| = \begin{cases} x & \text{if } x \geq 0 \\ -x & \text{if } x < 0 \end{cases}$$

This says that the absolute value symbols do one of two things:

1. If the quantity x inside the absolute value symbols is non-negative (greater than or equal to 0), then the absolute value symbols do nothing and can be omitted.

2. If the quantity x inside the absolute value symbols is negative, the absolute value symbols will negate the quantity (multiply it by –1), which will make the final result positive.

For example, $|6| = 6$ because the quantity inside the absolute value symbols, 6, is already non-negative, so the absolute value symbols do nothing. But $|-3| = -(-3) = 3$, because the –3 inside the absolute value symbols was negative, so the absolute value symbols negated it giving a positive result.

This may seem like a complicated way to do something simple, but it turns out to be the key to solving absolute value equations. The first example illustrates the reasoning on a very simple problem.

EXAMPLE 1

Solve: $|x| = 5$

___

Solution: There are two possibilities.

1. If the quantity inside the absolute value symbols, x, is non-negative ($x \geq 0$), then the absolute value symbols can be omitted and the equation simplifies to $x = 5$.

2. If the quantity inside the absolute value symbols, x, is negative ($x < 0$), then the absolute value symbols negate it and the equation becomes $-x = 5$, which is easily solved to give $x = -5$.

This equation appears to have two answers, $x = 5$ and $x = -5$. To be absolutely sure, we should check both answers in the original absolute value equation. We have $|5| = 5$ and $|-5| = 5$, so both answers check.

The solutions are $x = 5$ and $x = -5$.

___

Obviously, that was a very easy example. But it suggests a method that can be used to solve absolute value equations: rewrite each absolute value equation as two separate non-absolute value equations, solve them, and check the answers.

To solve an absolute value equation:

1. Write two separate non-absolute value equations:

    a. One the same as the original equation but with the absolute value symbols omitted.

    b. One where the entire quantity inside the absolute value symbols is negated.

2. Solve both equations.

3. Check the answers in the original equation (the one with the absolute value symbols). Reject any that don't check.

EXAMPLE 2

Solve: $|2x - 3| = 7$

Solution: 1. Write two separate non-absolute value equations.

$$2x - 3 = 7 \qquad \text{or} \qquad -(2x - 3) = 7$$

2. Solve both equations.

$$
\begin{array}{rcl}
2x - 3 = 7 & \text{or} & -(2x - 3) = 7 \\
2x = 10 & & -2x + 3 = 7 \\
x = 5 & & -2x = 4 \\
& & x = -2
\end{array}
$$

3. Check in the original equation.

$$
\begin{array}{rcl}
|2(5) - 3| = 7 & \qquad & |2(-2) - 3| = 7 \\
|7| = 7 & & |-7| = 7 \\
7 = 7 \ \checkmark & & 7 = 7 \ \checkmark
\end{array}
$$

The solution is $x = 5$ or $x = -2$.

The next example shows the importance of checking your candidate answers.

EXAMPLE 3

Solve: $|x - 2| - 3x = 6$

Solution: 1. Write two separate non-absolute value equations. Note that in the second equation, only the quantity inside the absolute value symbols, $x - 2$, is negated, not the $-3x$.

$$x - 2 - 3x = 6 \qquad \text{or} \qquad -(x - 2) - 3x = 6$$

2. Solve both equations.

$$
\begin{array}{rcl}
x - 2 - 3x = 6 & \text{or} & -(x - 2) - 3x = 6 \\
-2x = 8 & & -x + 2 - 3x = 6 \\
x = -4 & & -4x = 4 \\
& & x = -1
\end{array}
$$

3. Check in the original equation.

$$
\begin{array}{rcl}
|-4 - 2| - 3(-4) = 6 & \qquad & |-1 - 2| - 3(-1) = 7 \\
|-6| + 12 = 6 & & |-3| + 3 = 6 \\
18 = 6 \ \text{No} & & 6 = 6 \ \checkmark
\end{array}
$$

Only $x = -1$ checks. That is the only solution.

Try a couple yourself.

Problem 1

Solve: $|4x + 5| = 13$

Solution:

Problem 2

Solve: $|2x + 4| = x - 1$

Solution:

GRAPHING ABSOLUTE VALUE EQUATIONS IN TWO VARIABLES

In this book we will only graph absolute value equations in two variables that have the form $y = a|bx + c| + d$ where a, b, c, and d are constants and a and b are not both 0. Our approach will be much the same as it was for graphing quadratic equations in Unit 24. We will make a table using some well-chosen x-values.

EXAMPLE 4

Graph: $y = |2x - 4| - 3$

Solution: We will make a table of x-values. For this example, we'll use the integer x-values from $x = -1$ to $x = 5$. After this example, we'll discuss how to choose appropriate points to graph.

x	$y = \lvert 2x - 4 \rvert - 3$	(x, y)
−1	$y = \lvert 2(-1) - 4 \rvert - 3 = 3$	$(-1, 3)$
0	$y = \lvert 2(0) - 4 \rvert - 3 = 1$	$(0, 1)$
1	$y = \lvert 2(1) - 4 \rvert - 3 = -1$	$(1, -1)$
2	$y = \lvert 2(2) - 4 \rvert - 3 = -3$	$(2, -3)$
3	$y = \lvert 2(3) - 4 \rvert - 3 = -1$	$(3, -1)$
4	$y = \lvert 2(4) - 4 \rvert - 3 = 1$	$(4, 1)$
5	$y = \lvert 2(5) - 4 \rvert - 3 = 3$	$(5, 3)$

Now plot the points and connect them.

Answer:

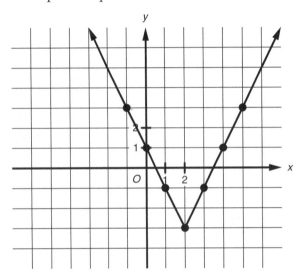

The graph of an absolute value equation in two variables is not a line, nor is it a parabola. Instead of the U-shape, ⌣, of a parabola, the graph of an absolute value equation is V-shaped: ⋁. It consists of two straight "half-lines" that meet at a corner called the vertex. The graphs of absolute values equations of the form $y = a\lvert bx + c \rvert + d$ all share some properties that are helpful to know when graphing them.

Properties of Absolute Value Graphs
$$y = a\lvert bx + c \rvert + d$$

1. The graph is ⋁-shaped, two half-lines that come to a point. (It is not rounded like a parabola, ⌣.)

2. If a is positive, the graph opens up: ⋁. If $a < 0$, the graph opens down: ⋀.

3. The graph will have a y-intercept, which can be found by substituting $x = 0$ into the equation and evaluating.

4. There are at most two x-intercepts, which can be found by substituting $y = 0$ into the equation and solving.

5. The graph has a vertical line of symmetry at $x = -\dfrac{c}{b}$.

6. The y-value of the vertex, which will be on the line of symmetry, is d.

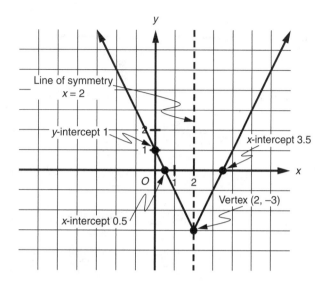

Based on the properties of absolute value graphs shown above, we suggest the following method for graphing equations of the form $y = a|bx + c| + d$.

Graphing an Absolute Value Equation in Two Variables
$$y = a|bx + c| + d$$

1. Write the equation in $y = a|bx + c| + d$ form.

2. Find the x-value of the line of symmetry: $x = -\dfrac{c}{b}$.

3. Make a table of values including the line of symmetry and at least one integral x-value on each side of the line of symmetry, a minimum of three points. Additional points on each side of the line of symmetry are helpful for checking the accuracy of hand-calculated and hand-drawn graphs.

4. Plot the points and connect them in a $\vee$ or $\wedge$ as appropriate.

5. Check that your graph matches the equation.

    • Is it $\vee$-shaped?

    • Does it open in the right direction, up if $a > 0$, down if $a < 0$?

    • Does the vertex have the correct y-value, d?

Steps 2 and 3 explain how we chose the x-values for Example 4. For $y = |2x - 4| - 3$, the line of symmetry is $x = -\dfrac{(-4)}{2}$, or $x = 2$. We wanted our table of values to include $x = 2$ and at least one integral x-value on each side. Thus we wanted to use at least $x = 1, 2,$ and 3. We included extra x-values on each side of the line of symmetry to ensure we got a good graph.

Let's see a second example.

EXAMPLE 5

Graph: $y = 3 - \dfrac{1}{2}|x|$

Solution: 1. Write the equation in $y = a|bx + c| + d$ form.

$y = -\dfrac{1}{2}|x| + 3$ or, to be very clear about the coefficients,

$y = -\dfrac{1}{2}|1x + 0| + 3$

2. Find the x-value of the line of symmetry.

$x = -\dfrac{c}{b} = -\dfrac{0}{1}$ or $x = 0$

3. Make a table of values including $x = 0$ and at least one x-value on each side. Because of the coefficient of $\dfrac{1}{2}$, it will be convenient to use even x-values.

x	$y = 3 - \dfrac{1}{2}|x|$	(x, y)
-4	$y = 3 - \dfrac{1}{2}|-4| = 1$	$(-4, 1)$
-2	$y = 3 - \dfrac{1}{2}|-2| = 2$	$(-2, 2)$
0	$y = 3 - \dfrac{1}{2}|0| = 3$	$(0, 3)$
2	$y = 3 - \dfrac{1}{2}|2| = 2$	$(2, 2)$
4	$y = 3 - \dfrac{1}{2}|4| = 1$	$(4, 1)$

4. Plot the points and connect them.

Answer:

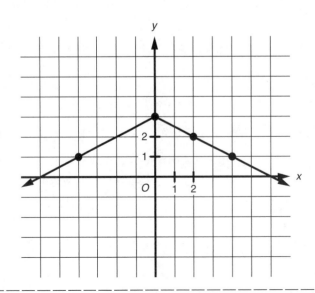

Now it's your turn.

Problem 3

Graph: $y = 2|x + 1| - 4$

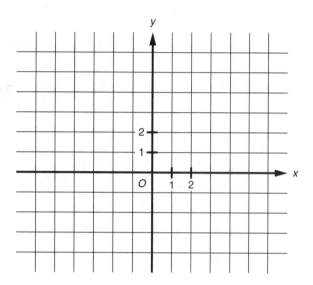

Problem 4

Graph: $y = -|3x - 6| + 4$

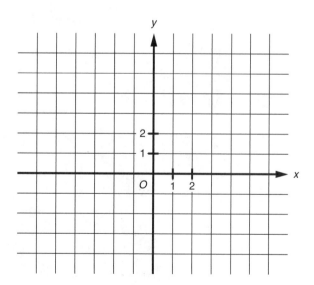

SOLVING ABSOLUTE VALUE INEQUALITIES

There are several methods for solving absolute value inequalities. The method we like isn't always the fastest, but it always works. It also has the advantage that it can be used for other kinds of inequalities, such as quadratic or higher order inequalities. This method is similar to the method used in Unit 29.

To solve an absolute value inequality:

1. First solve for equality. Change the inequality symbol to an equal sign and solve. Don't forget to check.

2. Graph the solutions from step 1 on a number line using appropriate circles: open for > or <, closed for ≥ or ≤. These solutions will divide your number line into up to three intervals.

3. Sketch a simple graph with the solutions from step 1 marked as the x-intercepts. Use the value of a to determine if the graph opens up or down. Determine from your sketch which values satisfy the original inequality and highlight that interval on your number line.

4. Write the solution set (the highlighted interval(s) on your number line) in appropriate form.

Some examples should make this clear.

EXAMPLE 6

Solve: $|3x - 4| < 10$

Solution: 1. Solve for equality.

$$|3x - 4| < 10$$

$$3x - 4 = 10 \qquad \text{or} \qquad -(3x - 4) = 10$$

$$3x = 14 \qquad\qquad\qquad -3x + 4 = 10$$

$$x = \frac{14}{3} \qquad\qquad\qquad -3x = 6$$

$$\qquad\qquad\qquad\qquad\qquad x = -2$$

Check:

$$\left|3\left(\frac{14}{3}\right) - 4\right| = 10 \qquad\qquad |3(-2) - 4| = 10$$

$$|10| = 10 \ \checkmark \qquad\qquad\qquad |-10| = 10 \ \checkmark$$

2. Put solutions on a number line. Use open circles because of the strict inequality.

3. Solving $|3x - 4| < 10$ is equivalent to solving $|3x - 4| - 10 < 0$. Sketch a simple graph for $y = |3x - 4| - 10$. The V-shaped graph will open up because $a > 0$. It will cross the x-axis at the solutions found in step 1, $x = -2$ and $x = \dfrac{14}{3}$.

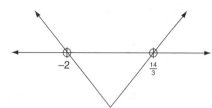

The solutions to the inequality $y < 0$ are the x-values where the graph is below the x-axis, which occurs on the interval $-2 < x < \dfrac{14}{3}$. On the number line, highlight the interval $-2 < x < \dfrac{14}{3}$.

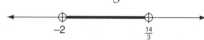

4. Write the solution.

The solution is all numbers between -2 and $\dfrac{14}{3}$.

Answer: $-2 < x < \dfrac{14}{3}$.

EXAMPLE 7

Solve: $\left|3 - \dfrac{1}{2}x\right| \geq 5$

Solution: 1. Solve for equality.

$$\left|3 - \frac{1}{2}x\right| = 5$$

$$3 - \frac{1}{2}x = 5 \quad \text{or} \quad -\left(3 - \frac{1}{2}x\right) = 5$$

$$-\frac{1}{2}x = 2 \qquad\qquad -3 + \frac{1}{2}x = 5$$

$$x = -4 \qquad\qquad\qquad \frac{1}{2}x = 8$$

$$x = 16$$

Check:

$$\left|3 - \frac{1}{2}(-4)\right| = 5 \qquad \left|3 - \frac{1}{2}(16)\right| = 5$$

$$|5| = 5 \ \checkmark \qquad\qquad |-5| = 5 \ \checkmark$$

2. Put solutions on a number line. Use closed circles because of the inclusive inequality (greater than *or equal to*).

3. Solving $\left|3 - \frac{1}{2}x\right| \geq 5$ is equivalent to solving $\left|3 - \frac{1}{2}x\right| - 5 \geq 0$. Sketch a simple graph for $\left|3 - \frac{1}{2}x\right| - 5$. The V-shaped graph will open up because $a > 0$. It will cross the x-axis at the solutions found in step 1, $x = -4$ and $x = 16$.

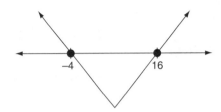

The solutions to the inequality $y \geq 0$ are the x-values where the graph is on or above the x-axis, which occurs on the intervals $x \leq -4$ or $x \geq 16$. On the number line, highlight the intervals $x < -4$ and $x > 16$.

4. Write the solution.

The solution is all numbers between less than or equal to –4 or greater than or equal to 16.

Answer: $x \leq -4$ or $x \geq 16$.

Problem 5

Solve: $|4x - 1| \leq 7$

Solution:

Problem 6

Solve: $|2x - 9| > 5$

Solution:

ALGEBRA AND THE CALCULATOR (Optional)

Find the absolute value function on your calculator. On newer TI-84s, you press
$\boxed{\text{ALPHA}}$ $\boxed{\text{WINDOW}}$ $\boxed{1}$.

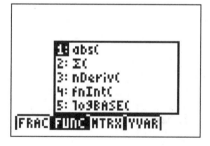

On older versions, you can find the absolute value function, abbreviated abs(, in the
Math menu; press $\boxed{\text{MATH}}$ $\boxed{)}$ $\boxed{1}$.

Absolute value equations are easy to graph on a graphing calculator. The key to an
accurate graph is to determine the x-value of the line of symmetry from either the table
or the formula before you graph the equation.

EXAMPLE 8

Graph: $y = -2|x + 3| + 4$

Solution: Type the equation into Y= to find the table of values and see the graph. Use the following keystrokes:

$\boxed{\text{Y=}}$ $\boxed{\text{CLEAR}}$ $\boxed{\text{(−)}}$ $\boxed{2}$ $\boxed{\text{ALPHA}}$ $\boxed{\text{WINDOW}}$ $\boxed{1}$

$\boxed{\text{X,T,θ,n}}$ $\boxed{+}$ $\boxed{3}$ $\boxed{)}$ $\boxed{+}$ $\boxed{4}$ $\boxed{\text{ENTER}}$

```
Plot1  Plot2  Plot3
\Y1 = -2|X+3|+4
\Y2 =
\Y3 =
\Y4 =
\Y5 =
\Y6 =
\Y7 =
```

To see the y-values, press $\boxed{\text{2nd}}$ $\boxed{\text{GRAPH}}$.

```
   X   │ Y1
   0   │ -2
   1   │ -4
   2   │ -6
   3   │ -8
   4   │ -10
   5   │ -12
   6   │ -14
 X=0
```

If the line of symmetry is an integer value, you will see a symmetric pattern in the y-values in the table. Scroll the table so that the row with $x = -3$ and $y = 4$ is in the middle of the table.

```
   X   │ Y1
  -6   │ -2
  -5   │ 0
  -4   │ 2
  -3   │ 4
  -2   │ 2
  -1   │ 0
   0   │ -2
 X=-6
```

Use these seven points to plot your graph. When you are done, check your graph by comparing it to the calculator's graph. The standard window $\boxed{\text{ZOOM}}$ $\boxed{6}$ is shown below.

Answer:

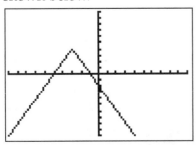

EXAMPLE 9

Graph: $y = |2x - 3| - 5$

Solution: Type the equation into Y= to find the table of values and see the graph. Use the following keystrokes .

| Y= | | CLEAR | | ALPHA | | WINDOW | | 1 | | 2 | | X,T,θ,n |

| − | | 3 | |) | | − | | 5 | | ENTER | .

```
Plot1  Plot2  Plot3
\Y1◻|2X-3|-5
\Y2=
\Y3=
\Y4=
\Y5=
\Y6=
\Y7=
```

To see the *y*-values, press | 2nd | | GRAPH | .

```
  X   │  Y1  │
──────┼──────┤
  0   │  -2  │
  1   │  -4  │
  2   │  -4  │
  3   │  -2  │
  4   │   0  │
  5   │   2  │
  6   │   4  │
──────┴──────┘
X=6
```

The line of symmetry is not an integer, so use the formula for the line of symmetry, $x = \dfrac{-c}{b}$, to find the value. For $y = |2x - 3| - 5$, $b = 2$, and $c = -3$, giving you

$$x = \frac{-(-3)}{2} = \frac{3}{2} = 1.5.$$

Look at the calculator's graph. The standard window | ZOOM | | 6 | is shown below.

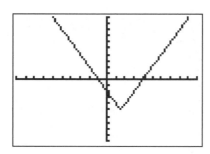

You need to graph the point at $x = 1.5$ and at least two points on each side. To find the vertex, (1.5, –5), press TRACE 1 · 5 ENTER.

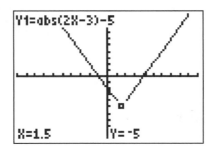

Use the table above for $x = 0, 1, 2,$ and 3.

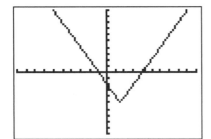

Answer:

_____

In this unit you learned to solve absolute value equations, graph absolute value equations, and solve absolute value inequalities.

To solve an absolute value equation:

1. Write two separate non-absolute value equations:

     a. One the same as the original equation but with the absolute value symbols omitted.

     b. One where the entire quantity inside the absolute value symbols is negated.

2. Solve both equations.

3. Check the answers in the original equation (the one with the absolute value symbols). Reject any that don't check.

To graph an absolute value equation of the form, $y = a|bx + c| + d$

1. Write the equation in $y = a|bx + c| + d$ form.

2. Find the x-value of the line of symmetry: $x = -\dfrac{c}{b}$.

3. Make a table of values including the line of symmetry and at least one integral x-value on each side of the line of symmetry, a minimum of three points. Additional points on each side of the line of symmetry are helpful for checking the accuracy of hand-calculated and hand-drawn graphs.

4. Plot the points and connect them in a ∨ or ∧ as appropriate.

5. Check that your graph matches the equation.

    • Is it ∨ -shaped?

    • Does it open in the right direction, up if $a > 0$, down if $a < 0$?

    • Does the vertex have the correct y-value, d?

To solve an absolute value inequality:

1. First solve for equality. Change the inequality symbol to an equal sign and solve. Don't forget to check.

2. Graph the solutions from step 1 on a number line using appropriate circles: open for > or <, closed for ≥ or ≤. These solutions will divide your number line into up to three intervals.

3. Sketch a simple graph with the solutions from step 1 marked as the x-intercepts. Use the value of a to determine if the graph opens up or down. Determine from your sketch which values satisfy the original inequality and highlight that interval on your number line.

4. Write the solution set (the highlighted interval(s) on your number line) in appropriate form.

EXERCISES

Solve the following equations algebraically.

1. $|2x - 5| = 21$

2. $|x - 7| + 2x = 5$

3. $|5 - 2x| = -3$

4. $3x - |x + 5| = 9$

Graph the following equations.

5. $y = |x - 4| - 2$

6. $y = -|x + 3| + 5$

7. $y = \frac{1}{2}|x - 2|$

8. $y = -|2x + 1| + 7$

Solve the following inequalities and sketch the solution on a number line.

9. $|3x - 7| \leq 8$

10. $|5 - x| > 3$

11. $|2x + 1| < 11$

12. $-2|x + 3| + 4 \geq 0$

Solutions to Problems

1. $|4x + 5| = 13$

$$4x + 5 = 13 \quad \text{or} \quad -(4x + 5) = 13$$

$$4x = 8 \qquad\qquad -4x - 5 = 13$$

$$x = 2 \qquad\qquad -4x = 18$$

$$x = -\frac{9}{2}$$

Both check. The solutions are $x = 2$ or $x = -4.5$.

2. $|2x + 4| = x - 1$

$2x + 4 = x - 1$ or $-(2x + 4) = x - 1$

$x = -5$ $-2x - 4 = x - 1$

$-3x = 3$

$x = -1$

Neither value checks. There is no solution.

3. Graph $y = 2|x + 1| - 4$

The x-value of the line of symmetry is $x = -1$.

Table of values

x	y
-3	0
-2	-2
-1	-4
0	-2
1	0

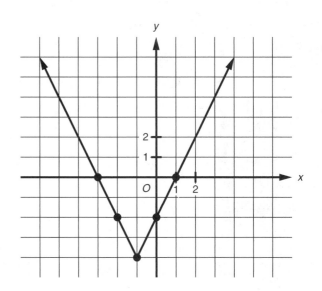

Answer:

4. Graph $y = -|3x - 6| + 4$

The x-value of the line of symmetry is $x = 2$.

Table of values

x	y
0	-2
1	1
2	4
3	1
4	-2

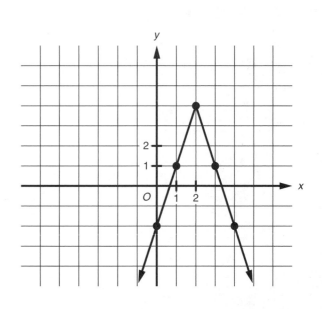

Answer:

5. $|4x - 1| \leq 7$

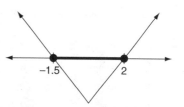

$$-(4x - 1) = 7$$

$$4x - 1 = 7 \qquad -4x + 1 = 7$$

$$4x = 8 \qquad -4x = 6$$

$$x = 2 \qquad \text{or} \qquad x = -1.5$$

Answer: $-1.5 \leq x \leq 2$

6. $|2x - 9| > 5$

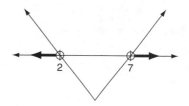

$$2x - 9 = 5 \qquad \text{or} \qquad -(2x - 9) = 5$$

$$2x = 14 \qquad -2x + 9 = 5$$

$$x = 7 \qquad -2x = -4$$

$$x = 2$$

Answer: $x < 2$ or $x > 7$

UNIT 31

Logarithms

The purpose of this unit is to provide you with a brief overview of logarithms. First the definition and notation used with logarithms will be introduced. Then you will learn about logarithmic properties and the ways these properties are used to simplify expressions involving logarithms.

The equation $\log_b N = x$ is read "the logarithm of N to the base b is x."

Definition: x is called the **logarithm of N to the base b** if $b^x = N$, where N and b are both positive numbers, $b \neq 1$.

In other words:

$$\log_b N = x \quad \text{if and only if} \quad b^x = N.$$

Examples of logarithms are:

$$\log_3 9 = 2 \quad \text{because} \quad 3^2 = 9$$

$$\log_2 8 = 3 \quad \text{because} \quad 2^3 = 8$$

$$\log_7 7 = 1 \quad \text{because} \quad 7^1 = 7$$

$$\log_5 25 = 2 \quad \text{because} \quad 5^2 = 25$$

$$\log_{13} 1 = 0 \quad \text{because} \quad 13^0 = 1$$

$$\log_2 \frac{1}{2} = -1 \quad \text{because} \quad 2^{-1} = \frac{1}{2}$$

LOGARITHMIC AND EXPONENTIAL FORMS

Notice that the **logarithm** of a positive number N is the **exponent** to which the base must be raised to produce the number N.

$$\log_b N = x \text{ is called the logarithmic form,}$$
$$b^x = N \text{ is called the exponential form,}$$
and the two statements are equivalent.

Try to develop your ability to go from one form to the other.

EXAMPLE 1

Express $\log_8 16 = \dfrac{4}{3}$ in exponential form.

Solution: If $\log_8 16 = \dfrac{4}{3}$,

then $8^{4/3} = 16$.

Express each of the following in exponential form.

Problem 1 $\log_3 243 = 5$

Problem 2 $\log_2 16 = 4$

Problem 3 $\log_{11} 121 = 2$

Problem 4 $\log_u w = y$

Students usually find it more difficult to go from exponential form to logarithmic form. We start by writing the base first, then the exponent. Remember that the logarithm of a number is an **exponent.**

EXAMPLE 2

Express $5^3 = 125$ in logarithmic form.

Solution: If $5^3 = 125$,

then $\log_5 125 = 3$.

Express each of the following in logarithmic form.

Problem 5 $4^2 = 16$

Problem 6 $12^0 = 1$

Problem 7 $4^{-5/2} = \dfrac{1}{32}$

Problem 8 $A^2 = c$

- -

Any positive number except 1 can be used for the base. Logarithms to the base 10 are called **common logs**, and the 10 is omitted from the logarithmic notation. In other words, if no base is written using logarithmic notation, the base is understood to be 10.

With common logs "the logarithm of N" is often shortened and read as "log N."

EXAMPLE 3

Express log $A = c$ in exponential form.

- -

Solution: log $A = c$ denotes a common logarithm with base 10 because no base number is indicated.

Therefore,   if   log $A = c$,
then $10^c = A$.

- -

It's your turn now.

Problem 9

Express log $100 = 2$ in exponential form.

- -

Solution:

- -

SOLVING LOGARITHMIC EQUATIONS

Now let's try solving some logarithmic equations of the form $x = \log_b N$, where we are to find the unknown number.

If you are more comfortable with the exponential form at this stage, we suggest changing from the logarithmic form to the exponential form to solve the equations. Eventually, though, you will need to solve the equations in their original form.

EXAMPLE 4

Solve: $\log_7 7 = x$.

Solution: If $\log_7 7 = x$,

then $7^x = 7$

and $7^x = 7^1$ since $7 = 7^1$.

Answer: $x = 1$

Here's an alternative method, which we think is easier:

In words, $\log_7 7 = x$ means "7 raised to what power x equals 7?"

The answer is 1.

EXAMPLE 5

Solve: $\log_2 32 = x$.

Solution: If $\log_2 32 = x$,

then $2^x = 32$

and $2^x = 2^5$ since $32 = 2^5$.

Answer: $x = 5$

Alternative method:

In words, $\log_2 32 = x$ means "2 raised to what power x equals 32?"

The answer is 5.

Try Problems 10–12 on your own.

Problem 10

Solve: $\log_8 64 = x$.

Solution:

Alternative method:

In words, $\log_8 64 = x$ means "8 raised to what power x equals 64?"

Problem 11

Solve: $\log 1000 = x$.

Solution:

Problem 12

Solve: $\log_7 1 = x$.

Solution:

In the preceding examples and problems, the base and number were given and we had to find the logarithm. The following examples illustrate the technique to be used when the log is given and we wish to solve for either the base or the number.

EXAMPLE 6

Solve: $\log_3 x = 4$.

Solution: If $\log_3 x = 4$,

        then $3^4 = x$

        and $81 = x$ since $3^4 = 81$.

Answer: $x = 81$

EXAMPLE 7

Solve: $\log_x 9 = \dfrac{1}{2}$.

Solution: If $\log_x 9 = \dfrac{1}{2}$,

        then $x^{1/2} = 9$

        or $\sqrt{x} = 9$.

Answer: $x = 81$

Since Example 7 may be a bit confusing, let's do one more before you try Problems 13–17.

EXAMPLE 8

Solve: $\log_x 4 = \dfrac{1}{3}$

Solution: If $\log_x 4 = \dfrac{1}{3}$

        then $x^{1/3} = 4$

        and $\sqrt[3]{x} = 4$

        since $\sqrt[3]{64} = 4$.

Answer: $x = 64$

Problem 13

Solve: $\log_5 x = 1$.

Solution:

Problem 14

Solve: $\log x = 4$.

Solution:

Problem 15

Solve: $\log_3 x = -2$.

Solution:

Problem 16

Solve: $\log_x 100 = 2$.

Solution:

Problem 17

Solve: $\log_x 8 = 3$.

Solution:

EXAMPLE 9

Solve: $\log 5 = x$.

Solution: Using a scientific or graphing calculator, you can find $\log 5 \approx 0.699$; however, you cannot find this value algebraically. The focus of this book is on algebraic methods, so problems like this one, which require a calculator, will not be addressed.

SIMPLIFYING EXPRESSIONS WITH LOGARITHMS

To accomplish simplification, we have **four** basic properties of logarithms.

These properties are used to shorten computations or to simplify complicated expressions involving products, quotients, powers, and roots.

Properties of Logarithms

Product property: $\quad \log_b AC = \log_b A + \log_b C$

Quotient property: $\quad \log_b \dfrac{A}{C} = \log_b A - \log_b C$

Power property: $\quad \log_b A^k = k \log_b A$

Root property: $\quad \log_b \sqrt[k]{A} = \dfrac{1}{k} \log A$

The following examples and problems illustrate how these four properties can be used to shorten computations. Remember that the logarithm of a positive number N is the exponent to which the base must be raised to produce the number N.

Product property: $\quad \log_b AC = \log_b A + \log_b C$

Verbally, the product property states that the logarithm of a product of two numbers is equal to the sum of the logarithms of the numbers.

EXAMPLE 10

Find: $\log_3(81 \cdot 9)$.

Solution: $\quad \log_3(81 \cdot 9) = \log_3 81 + \log_3 9$

$\qquad\qquad\qquad\quad = 4 + 2$

$\qquad\qquad\qquad\quad = 6$

Answer: $\quad \log_3(81 \cdot 9) = 6$

In case you are not convinced:

If $\quad \log_3(81 \cdot 9) \overset{?}{=} 6,$

then $\qquad\quad 3^6 \overset{?}{=} (81 \cdot 9)$

and $\qquad\quad 729 = 729.$

Problem 18

Find: $\log_2(8 \cdot 4)$.

Solution: Use the product property.

Problem 19

Find: $\log_2(64 \cdot 32)$.

Solution: Use the product property.

$$\boxed{\text{Quotient property:} \quad \log_b \frac{A}{C} = \log_b A - \log_b C}$$

In words, the quotient property states that the logarithm of a quotient of two numbers A and C is equal to the difference of the logarithms of the numbers A and C.

EXAMPLE 11

Find: $\log_3 \dfrac{1}{27}$.

Solution: $\log_3 \dfrac{1}{27} = \log_3 1 - \log_3 27$

$$= 0 - 3$$

$$= -3$$

Answer: $\log_3 \dfrac{1}{27} = -3$

If you are not convinced, change from the logarithmic form to the exponential form and verify that both sides represent the same number.

EXAMPLE 12

Question: If $\log 3 = 0.477$ and $\log 2 = 0.301$, what does $\log 1.5$ equal?

Solution: $\log 1.5 = \log\dfrac{3}{2}$

$= \log 3 - \log 2$

$= 0.477 - 0.301$

$= 0.176$

Answer: $\log 1.5 = 0.176$

If you are not convinced, use the log key on your calculator to verify that the log of 1.5 is 0.176.

Problem 20

Find: $\log_2 \dfrac{4}{64}$.

Solution: Use the quotient property.

Prroblem 21

Find: $\log_2 \dfrac{1}{4}$.

Solution: Use the quotient property.

$$\text{Power property:} \quad \log_b A^k = k \log_b A$$

In words, the power property states that the logarithm of a power of a number A is equal to the power times the logarithm of the number A.

EXAMPLE 13

Find: $\log_3(81)^5$.

Solution: $\log_3(81)^5 = 5\log_3 81$

$\qquad\qquad\qquad = 5(4)$

$\qquad\qquad\qquad = 20$

$$\text{Root property:} \quad \log_b \sqrt[k]{A} = \frac{1}{k}\log_b A$$

The root property follows directly from the power property. In words, the root property states that to take the logarithm of a root of a number A, rewrite the root of the number A using a fractional exponent and then apply the power property.

$$\log_b \sqrt[k]{A} = \log_b A^{\frac{1}{k}} = \frac{1}{k}\log_b A$$

The following example is an illustration of the use of this property.

EXAMPLE 14

Find: $\log_3 \sqrt{27}$.

Solution: $\log_3 \sqrt{27} = \log_3(27)^{1/2}$

$\qquad\qquad\qquad = \frac{1}{2}\log_3 27$

$\qquad\qquad\qquad = \frac{1}{2}(3)$

$\qquad\qquad\qquad = 1.5$

Problem 22

Find: $\log_2(32)^3$.

Solution: Use the power property.

Problem 23

Find: $\log_2 \sqrt{32}$.

Solution: Use the root property.

As previously stated, the four basic properties of logarithms are used also to expand a log of a complicated expression involving products, quotients, powers, and roots into simpler logs.

Let's try some more examples and problems. In each one, express the given logarithm in terms of logarithms of x, y, and z, where the variables are all positive.

EXAMPLE 15

Rewrite: $\log\left(\dfrac{xy}{z}\right)$.

Solution: $\log\left(\dfrac{xy}{z}\right) = \log(xy) - \log z$ quotient property

$$= \log x + \log y - \log z \quad \text{product property}$$

EXAMPLE 16

Rewrite: $\log\left(\dfrac{x^3y}{z^2}\right)$.

Solution: $\log\left(\dfrac{x^3y}{z^2}\right) = \log(x^3y) - \log z^2$ quotient property

$$= \log x^3 + \log y - \log z^2 \quad \text{product property}$$
$$= 3\log x + \log y - 2\log z \quad \text{power property}$$

EXAMPLE 17

Rewrite: $\log\sqrt{xy}$.

Solution: $\log\sqrt{xy} = \log(xy)^{1/2}$

$$= \frac{1}{2}\log(xy) \quad\quad \text{power property}$$

$$= \frac{1}{2}(\log x + \log y) \quad \text{product property}$$

$$= \frac{1}{2}\log x + \frac{1}{2}\log y$$

The last two problems are for you.

Problem 24

Rewrite: $\log(x^2y)$.

Solution:

Problem 25

Rewrite: $\log_a \sqrt{\dfrac{x}{y}}$.

Solution:

You should now have a basic understanding of logarithms. Remember that the logarithm of a positive number N is the exponent to which the base must be raised to produce the number N.

Also, you should be familiar with both logarithmic notation and exponential notation and be able to go from one form to the other.

Finally, you should be able to simplify logarithmic expressions involving products, quotients, and powers using the basic properties of logarithms.

Before beginning the next unit you should work the following exercises.

EXERCISES

Solve for x:

1. $\log_3 81 = x$ 6. $\log_7 x = 0$

2. $\log_5 125 = x$ 7. $\log_9 x = \dfrac{1}{2}$

3. $\log_7\left(\dfrac{1}{7}\right) = x$ 8. $\log_x 27 = 3$

4. $\log 1 = x$ 9. $\log_x 49 = 2$

5. $\log_3 x = 2$ 10. $\log_x 121 = 2$

Express each of the given logarithms in terms of logarithms of x, y, and z, where the variables are all positive.

11. $\log x^5$ 14. $\log \dfrac{x}{yz}$ Be careful!

12. $\log 2xy^3$ 15. $\log \sqrt{x^3 y}$

13. $\log \dfrac{x^2}{y}$

Solutions to Problems

1. $\log_3 243 = 5$
   $3^5 = 243$

2. $\log_2 16 = 4$
   $2^4 = 16$

3. $\log_{11} 121 = 2$
   $11^2 = 121$

4. $\log_u w = y$
   $u^y = w$

5. $4^2 = 16$
   $\log_4 16 = 2$

6. $12^0 = 1$
   $\log_{12} 1 = 0$

7. $4^{-5/2} = \dfrac{1}{32}$

   $\log_4 \dfrac{1}{32} = \dfrac{-5}{2}$

8. $A^2 = c$
   $\log_A c = 2$

9. $\log 100 = 2$
   $10^2 = 100$

10. $\log_8 64 = x$
    $8^x = 64$
    $8^2 = 64$
    $x = 2$

11. $\log 1000 = x$
    $10^x = 1000$
    $10^3 = 1000$
    $x = 3$

12. $\log_7 1 = x$
    $7^x = 1$
    $7^0 = 1$
    $x = 0$

13. $\log_5 x = 1$
    $5^1 = x$
    $5 = x$

14. $\log x = 4$
    $10^4 = x$
    $10000 = x$

15. $\log_3 x = -2$
    $3^{-2} = x$
    $\dfrac{1}{9} = x$

16. $\log_x 100 = 2$
    $x^2 = 100$
    $10^2 = 100$
    $x = 10$

17. $\log_x 8 = 3$
    $x^3 = 8$
    $2^3 = 8$
    $x = 2$

18. $\log_2 (8 \cdot 4) = \log_2 8 + \log_2 4$
    $= 3 + 4$
    $= 5$

19. $\log_2 (64 \cdot 32) = \log_2 64 + \log_2 32$
    $= 6 + 5$
    $= 11$

20. $\log_2 \left(\dfrac{4}{64} \right) = \log_2 4 - \log_2 64$
    $= 2 - 6$
    $= -4$

21. $\log_2 \left(\dfrac{1}{4} \right) = \log_2 1 - \log_2 4$
    $= 0 - 2$
    $= -2$

22. $\log_2 (32)^3 = 3 \log_2 32$
    $= 3(5)$
    $= 15$

23. $\log_2 \sqrt{32} = \log_2 (32)^{\frac{1}{2}}$
    $= \dfrac{1}{2} \log_2 32$
    $= \dfrac{1}{2}(5)$
    $= \dfrac{5}{2}$

24. $\log (x^2 y) = \log x^2 + \log y$
    $= 2 \log x + \log y$

25. $\log\sqrt{\dfrac{x}{y}} = \log\left(\dfrac{x}{y}\right)^{\frac{1}{2}}$

$\qquad\qquad = \dfrac{1}{2}\log\left(\dfrac{x}{y}\right)$

$\qquad\qquad = \dfrac{1}{2}(\log x - \log y)$

$\qquad\qquad = \dfrac{1}{2}\log x - \dfrac{1}{2}\log y$

UNIT 32

Right Triangles

The purpose of this, the last unit, is to provide you with a working knowledge of two frequently encountered triangles—the 30°-60°-90° triangle and the isosceles right triangle. When you have finished this unit, you will be able to recognize each of these triangles and, given the length of the one side, to find the lengths of the other two sides.

You probably recall that triangles are labeled using three capital letters, one at each vertex. The triangle shown below is referred to as triangle *ABC* or, simply, △*ABC*. The sides are labeled using lowercase letters as follows:

The side opposite angle *A* is *a*.
The side opposite angle *B* is *b*.
The side opposite angle *C* is *c*.

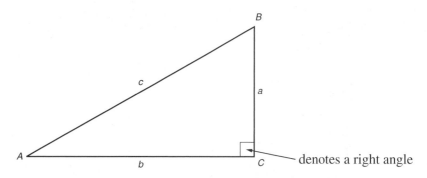

In any triangle the sum of the angles is equal to 180°. An angle of 90° is called a right angle. A **right triangle** is a triangle with a 90° angle. Thus triangle *ABC* is a right triangle.

In a right triangle the side opposite the right angle is called the **hypotenuse**. In triangle *ABC*, angle *C* is the right angle and *c* is the hypotenuse. The hypotenuse is longer than either of the other two sides, called legs.

One of the most useful theorems with regard to right triangles is the Pythagorean theorem. More than likely, you can recall some version of it yourself.

331

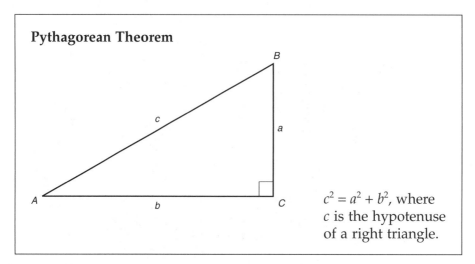

Pythagorean Theorem

$c^2 = a^2 + b^2$, where c is the hypotenuse of a right triangle.

Or, in words, given a right triangle, the square of the length of the hypotenuse is equal to the sum of the squared lengths of the sides (legs).

Be careful. The Pythagorean theorem is applicable only to right triangles.

EXAMPLE 1

Question: Given a right triangle with legs of 5 and 12, what is the length of the hypotenuse?

Solution: Begin by drawing a picture.

Label what is given in the problem.

Let $a = 5$ and $b = 12$.

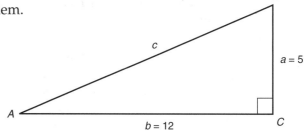

Because we are dealing with a right triangle, the Pythagorean theorem is applicable:

$$c^2 = a^2 + b^2$$
$$c^2 = (5)^2 + (12)^2$$
$$c^2 = 25 + 144$$
$$c^2 = 169$$
$$c = \sqrt{169}$$
$$c = 13$$

Answer: The length of the hypotenuse is 13.

Note: Only the positive square root of 169 is the answer, because c represents the length of a side of a triangle. A negative number would be meaningless for an answer in this situation.

EXAMPLE 2

Question: The hypotenuse of a right triangle is 3, and one leg is 2. What is the length of the third side?

- -

Solution: Begin by drawing a picture.

Label what is given in the problem.

Let $c = 3$ and $a = 2$.

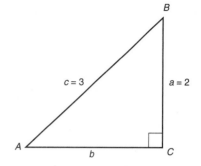

By the Pythagorean theorem:

$$c^2 = a^2 + b^2$$
$$(3)^2 = (2)^2 + b^2$$
$$9 = 4 + b^2$$
$$5 = b^2$$
$$\sqrt{5} = b$$

Answer: The length of the third side of the triangle is $\sqrt{5}$ or, if an approximation is requested, 2.236.

- -

We are sure you are ready to try a few problems now.

Problem 1

Question: One leg of a right triangle is 7, and the other leg is 5. What is the length of the hypotenuse?

- -

Solution:

Problem 2

Question: The hypotenuse of a right triangle is 5, and a leg is 3. What is the length of the third side?

- -

Solution:

- -

Notice that, when you are given the lengths of **any two sides of a right triangle**, the length of the third side can be found by using the Pythagorean theorem.

Problem 3

Question: The longest side of a right triangle is 11, and the shortest side is 3. What is the length of the other side?

- -

Solution:

- -

30°-60°-90° TRIANGLES

One of the most frequently encountered right triangles is the **30°-60°-90° triangle**, so named because

one angle is 30°,
another angle is 60°, and
the third angle is 90°.

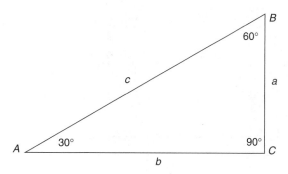

In a 30°-60°-90° triangle the sides have the following relationships:

The length of the **side opposite the 30° angle** is equal to half the length of the hypotenuse, or, stated as a formula,

$$a = \frac{c}{2}$$

The length of the **side opposite the 60° angle** is equal to half the length of the hypotenuse times $\sqrt{3}$, or, stated as a formula,

$$b = \frac{c}{2}\sqrt{3} \quad \text{or} \quad b = a\sqrt{3}$$

EXAMPLE 3

Question: If the hypotenuse of a 30°-60°-90° triangle is 6, what are the lengths of the other sides?

Solution: Begin by drawing a picture.

Label what is given in the problem.

Given $c = 6$.

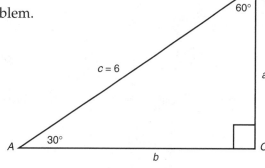

The length of the side opposite 30° is half the hypotenuse.

$$a = \frac{c}{2}$$

$$a = \frac{6}{2}$$

$$a = 3$$

The length of the side opposite 60° is half the hypotenuse times $\sqrt{3}$.

$$b = \frac{c}{2}\sqrt{3}$$

$$b = \frac{6}{2}\sqrt{3}$$

$$b = 3\sqrt{3}$$

Answer: $a = 3$ and $b = 3\sqrt{3}$, located as shown in the drawing

_ _

Try Problems 4–6.

Problem 4

Question: Given the drawing at the right, what are the lengths of a and b?

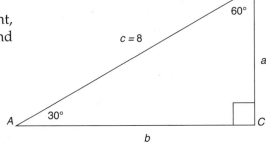

_ _

Solution:

_ _

Now suppose we change the conditions around a bit.

Problem 5

Question: If the length of the side opposite a 30° angle in a right triangle is 10, what is the length of the hypotenuse?

Solution:

Problem 6

Question: If the side opposite a 30° angle in a right triangle is 17, what is the length of the side opposite the 60° angle?

Solution:

In the next example we will do the more difficult situation where the length of the side opposite the 60° angle is given.

EXAMPLE 4

Question: If the side opposite the 60° angle in a right triangle is 8, what are the lengths of the other two sides?

Solution: We are given $b = 8$.

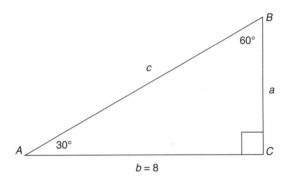

We know that the leg opposite the 60° angle is equal to the shorter leg times $\sqrt{3}$.

$$b = 8$$
$$b = a\sqrt{3}$$
$$8 = a\sqrt{3}$$
$$\frac{8}{\sqrt{3}} = a$$

Or, if you prefer the denominator rationalized:

$$a = \frac{8}{\sqrt{3}} \cdot \frac{\sqrt{3}}{\sqrt{3}} = \frac{8\sqrt{3}}{3}$$

Once the shorter leg is found, the hypotenuse is simply twice the shorter leg.

$$c = 2a$$

$$c = 2\left(\frac{8}{\sqrt{3}}\right) \qquad \text{or} \qquad c = 2\left(\frac{8\sqrt{3}}{3}\right)$$

$$c = \frac{16}{\sqrt{3}} \qquad \text{or} \qquad c = \frac{16\sqrt{3}}{3}$$

The shorter leg is $\frac{8}{\sqrt{3}}$ or $\frac{8\sqrt{3}}{3}$. The hypotenuse is $\frac{16}{\sqrt{3}}$ or $\frac{16\sqrt{3}}{3}$.

Here's a similar one for you.

Problem 7

Question: If the side opposite the 60° angle of a right triangle is 2, what are the lengths of the other two sides?

- -

Solution:

- -

ISOSCELES RIGHT TRIANGLES

A triangle with two equal sides is called an **isosceles triangle**.

Another frequently encountered right triangle is the **isosceles right triangle**, so named because

two sides are equal, and
the angle between them is 90°.

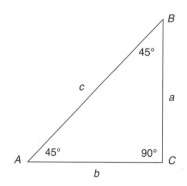

In an isosceles right triangle, if c is the hypotenuse, $a = b$ and angle A = angle B = 45°. The two equal angles of an isosceles triangle are often referred to as the **base angles**.

As in the 30°-60°-90° triangle, the sides of an **isosceles right triangle** are related.

In an isosceles right triangle the sides have the following relationships:

> The **sides opposite the 45° angles** are equal.
> Each side is equal to half the hypotenuse times $\sqrt{2}$, or, stated as a formula,
>
> $$a = b = \frac{c}{2}\sqrt{2}$$

EXAMPLE 5

Question: If an isosceles right triangle has a hypotenuse of 20, what are the lengths of the other sides?

Solution: As before, begin by drawing a picture.

Label what is given in the problem.

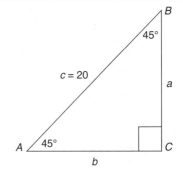

Because this is an isosceles right triangle, angle A and angle B are both 45° and $a = b$. Also, $c = 20$.

In an isosceles right triangle, the side opposite the 45° angle is half the hypotenuse times $\sqrt{2}$; therefore

$$b = \frac{c}{2}\sqrt{2}$$

and, by substitution,

$$b = \frac{20}{2}\sqrt{2}$$
$$= 10\sqrt{2}$$

Answer: The two equal sides are $10\sqrt{2}$.

This time you try the harder problem.

Problem 8

Question: If the side opposite a 45° angle in a right triangle is 5, what is the length of the
hypotenuse?

Solution:

Did you observe that, when working with either the 30°-60°-90° triangle or the isosceles
right triangle, if you are given only the length of one side it is possible to find the lengths
of the other two? In contrast, the Pythagorean theorem requires that two sides be known in
order to find the third.

The preceding examples have been fairly easy, as you probably guessed, and the problems that you are likely to encounter rarely appear in such a straightforward manner.
Instead, the information is contained in a word problem describing some real-life situation.
However, if you remember to draw a diagram and label what is given in the problem,
solving it should not be any more difficult than solving the problems you have done already.

A few word problems follow to illustrate what we mean.

EXAMPLE 6

A ladder leans against the side of the building with its foot 7 ft from the building. If the
ladder makes an angle of 60° with the ground, how long is the ladder?

Solution: Begin by drawing a diagram.

Label what is given in the problem.

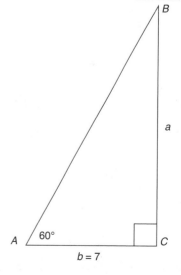

Angle $A = 60°$ and $b = 7$.

Therefore angle B must be $30°$. To find the length of the ladder, we apply our knowledge that the side opposite the $30°$ angle is half the hypotenuse. Since $b = 7$, c must be 14.

Answer: The ladder is 14 ft long.

EXAMPLE 7

When the sun is $45°$ above the horizon, how long is the shadow cast by a tree 40 ft high?

Solution: We are given angle $A = 45°$ and $a = 40$.

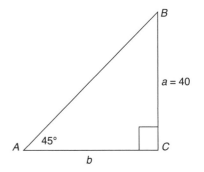

Since this is a right triangle, angle $B = 45°$ and we have an isosceles right triangle. Therefore b must be 40.

Answer: The shadow is 40 ft long.

EXAMPLE 8

From an observer on the ground, the angle of elevation to the top of a tree, 25 ft above the ground is $30°$. How far is the tree from the observer?

Solution: We will draw the diagram and label what is given.

$30°$ = angle of elevation

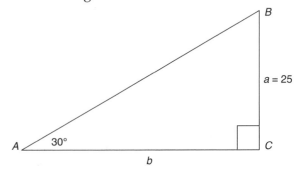

$$a = 25$$
$$b = a\sqrt{3}$$
$$b = 25\sqrt{3}$$

Answer: The tree is $25\sqrt{3}$ ft away.

Problem 9

Find the height of a street lamp if the angle of elevation of its top is 45° to an observer on the ground at a distance of 25 ft from its base.

Solution:

Problem 10

A man drives 1000 ft along a road that is inclined 30° to the horizontal. How high above his starting point is he?

Solution:

You probably did fine on that problem, but let's do one more.

Problem 11

A flagpole broken by the wind forms a right triangle with the ground. If the broken part makes an angle of 45° with the ground, and if the top of the flagpole is now 10 ft from its base, how tall was the flagpole?

—————————————————————————————————————

Solution:

—————————————————————————————————————

You should now be able to define and label a right triangle. Remember that, when you are given the lengths of any two sides of a right triangle, the length of the third side can be found by using the Pythagorean theorem, $c^2 = a^2 + b^2$, where c is the hypotenuse.

In addition you should be able to recognize two well-known right triangles.

1. **The 30°-60°-90° triangle** with properties:
    a. the side opposite the 30° angle is equal to one-half of the hypotenuse;
    b. the side opposite the 60° angle is equal to one-half of the hypotenuse times $\sqrt{3}$.

2. **The isosceles right triangle** with properties:
    a. the sides opposite the 45° angles are equal;
    b. the sides opposite the 45° angles are equal to one-half of the hypotenuse times $\sqrt{2}$.

Remember that, when working with either a 30°-60°-90° triangle or an isosceles right triangle, you need to be given only the length of one side in order to find the lengths of the other two sides. Also remember that the suggested procedure when solving such problems is to begin by drawing a diagram; then label what is given in the problem.

Congratulations; you have completed the book.

We can no longer say, "before continuing to the next unit," but do the following exercises anyway. There are only ten. Complete solutions are given at the end of the book for all ten exercises.

EXERCISES

Solve:

1. The hypotenuse of a right triangle is 7, and a leg is 3. Find the length of the third side.

2. Two legs of a right triangle are 1 and 2. Find the length of the hypotenuse.

3. If the hypotenuse of a 30°-60°-90° triangle is 10, find the lengths of the other sides.

4. If the side opposite the 60° angle in a right triangle is 5, what are the lengths of the other two sides?

5. If the side opposite the 30° angle in a right triangle is 1, what are the lengths of the other two sides?

6. If the side opposite a 45° angle in a right triangle is 6, what are the lengths of the other two sides?

7. If the hypotenuse of an isosceles right triangle is 10, what are the lengths of the other two sides?

8. A ladder 20 ft long is resting against a house. If it makes an angle of 45° with the ground, how high up does it reach?

9. An observer notes that the angle of elevation of the top of a neighboring building is 60°. If the distance from the observer to the neighboring building is 40 ft, how tall is the building?

10. When the sun is 30° above the horizon, an observer's shadow is 9 ft. How tall is the observer?

Solutions to Problems

1. $7^2 + 5^2 = c^2$

    $49 + 25 = c^2$

    $74 = c^2$

    $\sqrt{74} = c$

2. $3^2 + b^2 = 5^2$

    $9 + b^2 = 25$

    $b^2 = 16$

    $b = 4$

3. $3^2 + b^2 = 11^2$

    $9 + b^2 = 121$

$b^2 = 112$

$b = \sqrt{112}$

$b = \sqrt{16 \cdot 7}$

$b = 4\sqrt{7}$

4. $a = \dfrac{c}{2}$

    $a = \dfrac{8}{2}$

    $a = 4$

    $b = a\sqrt{3}$

    $b = 4\sqrt{3}$

5.

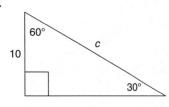

$a = 10$

$c = 2a$

$c = 2(10)$

$c = 20$

6.

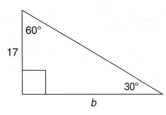

$a = 17$

$b = a\sqrt{3}$

$b = 17\sqrt{3}$

7.

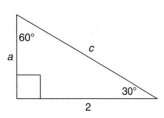

$b = 2$

$b = a\sqrt{3}$

$2 = a\sqrt{3}$

$\dfrac{2}{\sqrt{3}} = a$

$a = \dfrac{2}{\sqrt{3}} \cdot \dfrac{\sqrt{3}}{\sqrt{3}} = \dfrac{2\sqrt{3}}{3}$

$c = 2a$

$c = 2\left(\dfrac{2}{\sqrt{3}}\right)$ or $c = 2\left(\dfrac{2\sqrt{3}}{3}\right)$

$c = \dfrac{4}{\sqrt{3}}$ or $c = \dfrac{4\sqrt{3}}{3}$

8. $a = 5$

$a = \dfrac{c}{2}\sqrt{2}$

$5 = \dfrac{c}{2}\sqrt{2}$

$10 = c\sqrt{2}$

$\dfrac{10}{\sqrt{2}} = c$

$c = \dfrac{10}{\sqrt{2}} \cdot \dfrac{\sqrt{2}}{\sqrt{2}} = \dfrac{10\sqrt{2}}{2} = 5\sqrt{2}$

9.

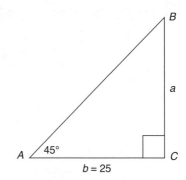

$b = 25$

$a = b$

$a = 25$

The street lamp is 25 feet high.

10.

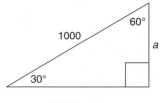

$c = 1000$

$c = 2a$

$1000 = 2a$

$500 = a$

He is 500 feet above his starting point.

11.

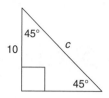

The flagpole is broken into two sides of the triangle. To find the total height of the flagpole, add the leg and the hypotenuse of the triangle.

$$a = 10$$

$$c = a\sqrt{2}$$

$$c = 10\sqrt{2}$$

$$c + a = 10 + 10\sqrt{2}$$

The flagpole was $10 + 10\sqrt{2}$ feet tall.

Answers to Exercises

UNIT 1

1. -7
2. 4
3. -9
4. -1
5. -45
6. -18
7. 24
8. -4
9. 5
10. 1
11. -1
12. -3
13. 9
14. -3
15. -6
16. -10
17. -24
18. 32
19. 0
20. 1
21. $\begin{aligned} 1 + 6 \cdot 3 \div 2 - 2 &= 1 + 18 \div 2 - 2 \\ &= 1 + 9 - 2 \\ &= 10 - 2 \\ &= 8 \end{aligned}$

22. $\begin{aligned} 12 \div 3 \cdot 2 - 1 &= 4 \cdot 2 - 1 \\ &= 8 - 1 \\ &= 7 \end{aligned}$

23. $\begin{aligned} 0 \cdot 7 + 15 \div 5 - 2 \cdot 1 &= 0 + 3 - 2 \\ &= 1 \end{aligned}$

24. $\begin{aligned} (-10) \cdot 12 - 6 \cdot 3 &= -120 - 18 \\ &= -138 \end{aligned}$

25. $\begin{aligned} |-5| + |3| - |-2| &= 5 + 3 - 2 \\ &= 8 - 2 \\ &= 6 \end{aligned}$

26. $\begin{aligned} -|-3| + |7| &= -3 + 7 \\ &= 4 \end{aligned}$

27. $\begin{aligned} -6(7 - 4) + 2(3 - 5) &= -6(3) + 2(-2) \\ &= -18 + -4 \\ &= -22 \end{aligned}$

28. $\begin{aligned} 7(3 - 6) - 8(4 - 1) &= 7(-3) - 8(3) \\ &= -21 - 24 \\ &= -45 \end{aligned}$

UNIT 2

1. $-7(x + 2y - 3)$

   $-7x - 14y + 21$

2. $3x + 4x + (x + 2)$

   $3x + 4x + x + 2$

   $8x + 2$

3. $2(x + y) + 7x + 3$

   $2x + 2y + 7x + 3$

   $9x + 2y + 3$

4. $-1(-2x + 1) + 1$

   $2x - 1 + 1$

   $2x$

5. $7x - 2y + 5 + (2x + 5y - 4)$
   $7x - 2y + 5 + 2x + 5y - 4$
   $9x + 3y + 1$

6. $(3x + 5xy + 2y) + (4 - 3xy - 2x)$
   $3x + 5xy + 2y + 4 - 3xy - 2x$
   $x + 2xy + 2y + 4$

7. $(5y - 2a + 1) + 2(y - 3a - 7)$
   $5y - 2a + 1 + 2y - 6a - 14$
   $7y - 8a - 13$

8. $3(2a - b) - 4(b - 2a)$
   $6a - 3b - 4b + 8a$
   $14a - 7b$

9. $2(7x - 5 + y) - (y + 7)$
   $2(7x - 5 + y) - 1(y + 7)$
   $14x - 10 + 2y - y - 7$
   $14x + y - 17$

10. $5 - 2[x + 2(3 + x)]$
    $5 - 2[x + 6 + 2x]$
    $5 - 2x - 12 - 4x$
    $-6x - 7$

11. $x - [2x - 2(1 - x)]$
    $x - [2x - 2 + 2x]$
    $x - 1[4x - 2]$
    $x - 4x + 2$
    $-3x + 2$

12. $4 - 9(2x - 3) + 7(x - 1)$
    $4 - 18x + 27 + 7x - 7$
    $24 - 11x$

13. $3(x + 4) + 5x - 8$
    $3x + 12 + 5x - 8$
    $8x + 4$

14. $4x - [5 - 3(2x - 6)]$
    $4x - 1[5 - 6x + 18]$
    $4x - 1[23 - 6x]$
    $4x - 23 + 6x$
    $10x - 23$

UNIT 3

1. $2x - 7 = 9 - 6x$

   $2x + 6x = 9 + 7$
   $8x = 16$
   $\dfrac{\cancel{8}x}{\cancel{8}} = \dfrac{\cancel{16}^{2}}{8}$
   $x = 2$

2. $2(x + 1) - 3(4x - 2) = 6x$

   $2x + 2 - 12x + 6 = 6x$
   $-10x + 8 = 6x$
   $8 = 6x + 10x$
   $8 = 16x$
   $\dfrac{8}{16} = \dfrac{\cancel{16}x}{\cancel{16}}$
   $\dfrac{1}{2} = x$

3.

$$\frac{x-3}{4} = 5$$

$$4\left(\frac{x-3}{4} = 5\right)$$

$$4\left(\frac{x-3}{4}\right) = 5 \cdot 4$$

$$x - 3 = 20$$

$$x = 20 + 3$$

$$x = 23$$

4.

$$20 - \frac{3x}{5} = x - 12$$

$$5\left(20 - \frac{3x}{5} = x - 12\right)$$

$$100 - \frac{5 \cdot 3x}{5} = 5x - 60$$

$$100 - 3x = 5x - 60$$

$$-3x - 5x = -60 - 100$$

$$-8x = -160$$

$$\frac{-8x}{-8} = \frac{-160}{-8}$$

$$x = 20$$

5.

$$2x - 9 = 5x - 15$$

$$2x - 5x = -15 + 9$$

$$-3x = -6$$

$$\frac{-3x}{-3} = \frac{-6^2}{-3}$$

$$x = 2$$

6.

$$2(x+2) = 5 + \frac{x+1}{3}$$

$$3\left(2(x+2) = 5 + \frac{x+1}{3}\right)$$

$$3(2(x+2)) = 3(5) + 3\left(\frac{x+1}{3}\right)$$

$$6(x+2) = 15 + x + 1$$

$$6x + 12 = 15 + x + 1$$

$$6x - x = 16 - 12$$

$$5x = 4$$

$$\frac{5x}{5} = \frac{4}{5}$$

$$x = \frac{4}{5}$$

7.

$$15 - 3(9 - x) = x$$

$$15 - 27 + 3x = x$$

$$-12 + 3x = x$$

$$3x - x = 12$$

$$2x = 12$$

$$\frac{2x}{2} = \frac{12^6}{2}$$

$$x = 6$$

8. $$3 - \frac{5(x-1)}{2} = x$$

$$2\left(3 - \frac{5(x-1)}{2} = x\right)$$

$$6 - \frac{2 \cdot 5(x-1)}{2} = 2x$$

$$6 - 5(x-1) = 2x$$

$$6 - 5x + 5 = 2x$$

$$11 - 5x = 2x$$

$$11 = 2x + 5x$$

$$11 = 7x$$

$$\frac{11}{7} = \frac{7x}{7}$$

$$\frac{11}{7} = x$$

9. $$x - \frac{x-1}{4} = 0$$

$$4\left(x - \frac{x-1}{4} = 0\right)$$

$$4x - \frac{4(x-1)}{4} = 0$$

$$4x - 1(x-1) = 0$$

$$4x - x + 1 = 0$$

$$3x + 1 = 0$$

$$3x = 0 - 1$$

$$3x = -1$$

$$\frac{3x}{3} = \frac{-1}{3}$$

$$x = \frac{-1}{3}$$

10. $$1 - \frac{x}{2} = 5$$

$$2\left(1 - \frac{x}{2} = 5\right)$$

$$2 - \frac{2x}{2} = 10$$

$$2 - x = 10$$

$$-x = 10 - 2$$

$$-x = 8$$

$$x = -8$$

11. $$-5w - 1 = -9w - 1$$

$$-5w + 9w = -1 + 1$$

$$4w = 0$$

$$\frac{4w}{4} = \frac{0}{4}$$

$$w = 0$$

12. $$3(-2x + 1) = -6x - 7$$

$$-6x + 3 = -6x - 7$$

$$-6x + 6x = -7 - 3$$

$$0 = -10$$

But 0 does not equal −10; therefore the problem has no solution.

13. $$2(4a + 1) = 4(2a - 1) + 6$$

$$8a + 2 = 8a - 4 + 6$$

$$8a + 2 = 8a + 2$$

An identity, and the solution is all real numbers.

14. $3(z + 5) - 2z = \dfrac{z - 1}{2} + 17$

$2\left(3(z + 5) - 2z = \dfrac{z - 1}{2} + 17\right)$

$6(z + 5) - 4z = \dfrac{2(z - 1)}{2} + 34$

$6z + 30 - 4z = z - 1 + 34$

$2z + 30 = z + 33$

$2z - z = 33 - 30$

$z = 3$

15. $13 - (2y + 2) = 2(y + 2) + 3y$

$13 - 1(2y + 2) = 2(y + 2) + 3y$

$13 - 2y - 2 = 2y + 4 + 3y$

$11 - 2y = 5y + 4$

$11 - 4 = 5y + 2y$

$7 = 7y$

$\dfrac{7^1}{7} = \dfrac{7y}{7}$

$1 = y$

16. $3x + 4(x - 2) = x - 5 + 3(2x - 1)$

$3x + 4x - 8 = x - 5 + 6x - 3$

$7x - 8 = 7x - 8$

$7x - 7x = 8 - 8$

$0 = 0$

An identity, and the solution is all real numbers.

17. $2c + 3(c + 2) = 5c + 11$

$2c + 3c + 6 = 5c + 11$

$5c + 6 = 5c + 11$

$5c - 5c = 11 - 6$

$0 = 5$

No solution

18. $2[2 - x - (2x - 5)] = 11 - x$

$2[2 - x - 2x + 5] = 11 - x$

$2[7 - 3x] = 11 - x$

$14 - 6x = 11 - x$

$14 - 11 = 6x - x$

$3 = 5x$

$\dfrac{3}{5} = \dfrac{5x}{5}$

$\dfrac{3}{5} = x$

UNIT 4

1. $1 = \dfrac{5}{x}$

$\dfrac{1}{1} = \dfrac{5}{x}$

$1 \cdot x = 5 \cdot 1$

$x = 5$

Check: $1 = \dfrac{5}{5}$

$1 = 1$

2. $$\frac{x-3}{2} = \frac{2x+4}{5}$$

$5(x-3) = 2(2x+4)$ Cross-multiply.

$5x - 15 = 4x + 8$

$5x - 4x = 8 + 15$

$x = 23$

It is not necessary to check this, as we did not have a fractional equation—the variable was not in any denominator.

3. $$\frac{6}{x-2} = -3$$

$$\frac{6}{x-2} = \frac{-3}{1}$$

$6 \cdot 1 = -3(x-2)$

$ 6 = -3x + 6$

$6 - 6 = -3x$

$ 0 = -3x$

$$\frac{0}{-3} = \frac{\cancel{-3}x}{\cancel{-3}}$$

$\phantom{\frac{0}{-3}} 0 = x$

Check: $\dfrac{6}{0-2} = -3$

$ \dfrac{6}{-2} = -3$

$ -3 = -3$

4. $$\frac{3x-3}{x-1} = 2$$

$3x - 3 = 2(x-1)$ Cross-multiply.

$3x - 3 = 2x - 2$

$3x - 2x = -2 + 3$

$ x = 1$

Check: $\dfrac{3-3}{1-1} \overset{?}{=} 2$

But we cannot have 0 in the denominator; therefore there is no solution to this equation.

5.
$$\frac{x}{2} = \frac{x+6}{3}$$
$$3x = 2(x+6)$$
$$3x = 2x + 12$$
$$3x - 2x = 12$$
$$x = 12$$

Check:
$$\frac{12}{2} = \frac{12+6}{3}$$
$$6 = \frac{18}{3}$$
$$6 = 6$$

6.
$$\frac{3}{x} = \frac{4}{x-2}$$
$$3(x-2) = 4x$$
$$3x - 6 = 4x$$
$$3x - 4x = 6$$
$$-x = 6$$
$$x = -6$$

Check:
$$\frac{3}{x} = \frac{4}{x-2}$$
$$\frac{3}{(-6)} = \frac{4}{(-6)-2}$$
$$-\frac{1}{2} = -\frac{1}{2}$$

7.
$$\frac{4}{x+3} = \frac{1}{x-3}$$
$$4(x-3) = x+3$$
$$4x - 12 = x + 3$$
$$4x - x = 3 + 12$$
$$3x = 15$$
$$\frac{\cancel{3}x}{\cancel{3}} = \frac{\cancel{15}^{5}}{\cancel{3}}$$
$$x = 5$$

Check:
$$\frac{4}{5+3} = \frac{1}{5-3}$$
$$\frac{4}{8} = \frac{1}{2}$$
$$\frac{1}{2} = \frac{1}{2}$$

8. $$\frac{5-2x}{x-1} = -2$$

$$\frac{5-2x}{x-1} = \frac{-2}{1}$$

$$5 - 2x = -2(x-1)$$

$$5 - 2x = -2x + 2$$

$$-2x + 2x = 2 - 5$$

$$0 = -3$$

No solution, since 0 does not equal −3.

9. $$\frac{x+3}{x-2} = 2$$

$$\frac{x+3}{x-2} = \frac{2}{1}$$

$$2(x-2) = x+3$$

$$2x - 4 = x + 3$$

$$2x - x = 3 + 4$$

$$x = 7$$

Check: $$\frac{7+3}{7-2} = 2$$

$$\frac{10}{5} = 2$$

$$2 = 2$$

10. $$\frac{4}{5}x - \frac{1}{4} = -\frac{3}{2}x$$

$$20\left(\frac{4}{5}x - \frac{1}{4} = -\frac{3}{2}x\right)$$

$$\frac{\overset{4}{\cancel{20}} \cdot 4 \cdot x}{\cancel{5}} - \frac{\overset{5}{\cancel{20}} \cdot 1}{\cancel{4}} = -\frac{\overset{10}{\cancel{20}} \cdot 3 \cdot x}{\cancel{2}}$$

$$16x - 5 = -30x$$

$$16x + 30x = 5$$

$$46x = 5$$

$$\frac{\cancel{46}x}{\cancel{46}} = \frac{5}{46}$$

$$x = \frac{5}{46}$$

The variable was not in any denominator; therefore it is not necessary to check. The solution is $x = \dfrac{5}{46}$.

11.
$$5 + \frac{3+x}{x} = \frac{5}{x}$$

$$x\left(5 + \frac{3+x}{x} = \frac{5}{x}\right)$$

$$5x + \frac{\cancel{x}(3+x)}{\cancel{x}} = \frac{5\cancel{x}}{\cancel{x}}$$

$$5x + 3 + x = 5$$

$$6x + 3 = 5$$

$$6x = 5 - 3$$

$$6x = 2$$

$$\frac{\cancel{6}x}{\cancel{6}} = \frac{2^1}{\cancel{6}_3}$$

$$x = \frac{1}{3}$$

Check: The check has a lot of fractions and could take a long time to evaluate. Look at the original equation. The denominators have just x; only $x = 0$ will make the equation undefined, so $x = \dfrac{1}{3}$ will check.

12.
$$\frac{4}{x-2} - \frac{1}{x} = \frac{5}{x-2}$$

$$\left(\frac{4}{x-2} - \frac{1}{x} = \frac{5}{x-2}\right)x(x-2)$$

$$\left(\frac{4}{x-2} - \frac{1}{x} = \frac{5}{x-2}\right)x(x-2)$$

$$\frac{4 \cdot x(\cancel{x-2})}{\cancel{x-2}} - \frac{1 \cdot \cancel{x}(x-2)}{\cancel{x}} = \frac{5 \cdot x(\cancel{x-2})}{\cancel{x-2}}$$

$$4x - (x-2) = 5x$$

$$4x - x + 2 = 5x$$

$$3x + 2 = 5x$$

$$2 = 5x - 3x$$

$$2 = 2x$$

$$\frac{2}{2} = \frac{\cancel{2}x}{\cancel{2}}$$

$$x = 1$$

Check:
$$\frac{4}{x-2} - \frac{1}{x} = \frac{5}{x-2}$$

$$\frac{4}{1-2} - \frac{1}{1} = \frac{5}{1-2}$$

$$\frac{4}{-1} - 1 = \frac{5}{-1}$$

$$-4 - 1 = -5$$

$$-5 = -5$$

UNIT 5

1. $\dfrac{2ax}{3c} = \dfrac{y}{m}$

   $2amx = 3cy$

   $\dfrac{2a\,mx}{2\,mx} = \dfrac{3cy}{2mx}$

   $a = \dfrac{3cy}{2mx}$

2. $2cy + 4d = 3ax - 4b$

   $2cy + 4d + 4b = 3ax$

   $\dfrac{2cy + 4d + 4b}{3x} = \dfrac{3ax}{3x}$

   $\dfrac{2cy + 4d + 4b}{3x} = a$

3. $ax + 3a = bx + 7c$

   $a(x + 3) = bx + 7c$

   $\dfrac{a(x + 3)}{x + 3} = \dfrac{bx + 7c}{x + 3}$

   $a = \dfrac{bx + 7c}{x + 3}$

4. $4x + 5c - 2a = 0$

   $4x + 5c = 2a$

   $\dfrac{4x + 5c}{2} = \dfrac{2a}{2}$

   $\dfrac{4x + 5c}{2} = a$

5. $a(x + 2) = \pi - cy$

   $\dfrac{a(x + 2)}{x + 2} = \dfrac{\pi - cy}{x + 2}$

   $a = \dfrac{\pi - cy}{x + 2}$

6. $\dfrac{2ax}{3c} = \dfrac{y}{m}$

   $2amx = 3cy$

   $\dfrac{2amx}{2am} = \dfrac{3cy}{2am}$

   $x = \dfrac{3cy}{2am}$

7. $2cy + 4d = 3ax - 4b$

   $2cy + 4d + 4b = 3ax$

   $\dfrac{2cy + 4d + 4b}{3a} = \dfrac{3ax}{3a}$

   $\dfrac{2cy + 4d + 4b}{3a} = x$

8. $ax + 3a = bx + 7c$

   $ax - bx = 7c - 3a$

   $x(a - b) = 7c - 3a$

   $\dfrac{x(a - b)}{a - b} = \dfrac{7c - 3a}{a - b}$

   $x = \dfrac{7c - 3a}{a - b}$

9. $4x + 5c - 2a = 0$

   $4x = 2a - 5c$

   $\dfrac{4x}{4} = \dfrac{2a - 5c}{4}$

   $x = \dfrac{2a - 5c}{4}$

10. $a(x + 2) = \pi - cy$

    $ax + 2a = \pi - cy$

    $ax = \pi - cy - 2a$

    $\dfrac{ax}{a} = \dfrac{\pi - cy - 2a}{a}$

    $x = \dfrac{\pi - cy - 2a}{a}$

11. $\dfrac{2ax}{3c} = \dfrac{y}{m}$

    $2amx = 3cy$

    $\dfrac{2amx}{3y} = \dfrac{3cy}{3y}$

    $\dfrac{2amx}{3y} = c$

12. $2cy + 4d = 3ax - 4b$

    $2cy = 3ax - 4b - 4d$

    $\dfrac{2cy}{2y} = \dfrac{3ax - 4b - 4d}{2y}$

    $c = \dfrac{3ax - 4b - 4d}{2y}$

13.
$$ax + 3a = bx + 7c$$
$$ax + 3a - bx = 7c$$
$$\frac{ax + 3a - bx}{7} = \frac{7c}{7}$$
$$\frac{ax + 3a - bx}{7} = c$$

14. $4x + 5c - 2a = 0$
$$5c = 2a - 4x$$
$$\frac{\cancel{5}c}{\cancel{5}} = \frac{2a - 4x}{5}$$
$$c = \frac{2a - 4x}{5}$$

15. $a(x + 2) = \pi - cy$
$$ax + 2a = \pi - cy$$
$$cy = \pi - ax - 2a$$
$$\frac{c\cancel{y}}{\cancel{y}} = \frac{\pi - ax - 2a}{y}$$
$$c = \frac{\pi - ax - 2a}{y}$$

16.
$$C = 2\pi r$$
$$\frac{C}{2\pi} = \frac{2\pi r}{2\pi}$$
$$\frac{C}{2\pi} = r$$

17.
$$R = \frac{V}{I}$$
$$\frac{R}{1} = \frac{V}{I}$$
$$IR = V$$
$$\frac{I\cancel{R}}{\cancel{R}} = \frac{V}{R}$$
$$I = \frac{V}{R}$$

18.
$$A = \frac{1}{2}(a + b)h$$
$$2A = \cancel{2} \cdot \frac{1}{\cancel{2}}(a + b)h$$
$$2A = (a + b)h$$
$$2A = ah + bh$$
$$2A - bh = ah$$
$$\frac{2A - bh}{h} = \frac{a\cancel{h}}{\cancel{h}}$$
$$\frac{2A - bh}{h} = a$$

UNIT 6

1. $x - 5$

2. $3x + 8$

3. $8x - 10$

4. $\dfrac{x}{3}$

5. $2x - 5 = 11$

6. $7x = 35$

7. $x - 20 = 32$

8. $x + 12 = 20$

9. $15 + 2x = 47$

10. $4 + 3x = 17$

11. Let x represent Jack's age.

    Then George's age is $x + 8$ because he is 8 years older than Jack.

    The sum of their ages is 42.

    Jack's age + George's age = 42

$$x \;+\; x + 8 = 42$$
$$2x + 8 = 42$$
$$2x = 42 - 8$$
$$2x = 34$$
$$\frac{\cancel{2}x}{\cancel{2}} = \frac{\cancel{34}^{17}}{\cancel{2}}$$
$$x = 17$$

    Jack is 17 and George is 25.

12. Let x = length of shorter piece of rope.

    Then $x + 10$ = length of longer piece.

    The length of the rope is 36 feet.

$$x + (x + 10) = 36$$
$$2x + 10 = 36$$
$$2x = 26$$
$$x = 13$$
$$x + 10 = 23$$

    The shorter piece of rope is 13 feet long, and the other is 23 feet long.

13. Let x represent the first ticket number.

    Then the next three ticket numbers are $x + 1$, $x + 2$, and $x + 3$ because they are consecutive numbers.

    The sum of the ticket numbers is 1354.

$$x + (x + 1) + (x + 2) + (x + 3) = 1354$$
$$4x + 6 = 1354$$
$$4x = 1354 - 6$$
$$4x = 1348$$
$$\frac{\cancel{4}x}{\cancel{4}} = \frac{\cancel{1348}^{337}}{\cancel{4}}$$
$$x = 337$$
$$x + 1 = 338$$
$$x + 2 = 339$$
$$x + 3 = 340$$

    The numbers on the raffle tickets are 337, 338, 339, and 340.

14. Let x represent the first even integer.

Then $x + 2$ represents the second consecutive even integer.

And $x + 4$ represents the third consecutive even integer.

The sum of three consecutive even integers is 6480.

$$x + (x + 2) + (x + 4) = 6480$$
$$3x + 6 = 6480$$
$$3x = 6474$$
$$x = 2158$$
$$x + 2 = 2160$$
$$x + 4 = 2162$$

The three consecutive even integers are 2158, 2160, and 2162.

15. Let w represent the width of the court.

Then $l = w + 34$ because the length is 34 ft longer than the width.

The perimeter is 268.

$$P = 2l + 2w$$
$$268 = 2(w + 34) + 2w$$
$$268 = 2w + 68 + 2w$$
$$268 = 4w + 68$$
$$268 - 68 = 4w$$
$$200 = 4w$$
$$\frac{\cancel{200}^{\,50}}{\cancel{4}} = \frac{\cancel{4}w}{\cancel{4}}$$
$$50 = w$$
$$l = w + 24$$
$$l = 50 + 34$$
$$l = 84$$

The dimensions of the basketball court are 50 ft by 84 ft.

16. Let x represent the width of the parking lot.

Then $3x$ represents the length since it is 3 times the width.

The perimeter is given at 4100 ft.

$$P = 2l + 2w$$
$$4100 = 2(3x) + 2(x)$$
$$4100 = 6x + 2x$$
$$4100 = 8x$$
$$512.5 = x$$
$$3x = 1537.5$$

The width of the parking lot is 512.5 ft, and the length is 1537.5 ft.

17. Let x represent Aurora's walking time. Chris walks for the same amount of time. The completed chart is:

	rate	×	time	=	distance
Aurora Chris	3 mph 2 mph		x x		$3x$ $2x$

In this problem, they live 4 miles apart, so their distances must add to 4 miles.

$$3x + 2x = 4$$
$$5x = 4$$
$$\frac{\cancel{5}x}{\cancel{5}} = \frac{4}{5}$$
$$x = \frac{4}{5}$$

Aurora and Chris will meet in $\frac{4}{5}$ hour or in 48 minutes.

18. Let x represent Rosalie's running time.
Then Sally's running time will be $x - 0.5$ because she started running $\frac{1}{2}$ hour later. The completed chart is:

	rate	×	time	=	distance
Rosalie Sally	6 mph 8 mph		x $x - 0.5$		$6x$ $8(x - 0.5)$

In this problem the distances run by each person are equal.

$$6x = 8(x - 0.5)$$
$$6x = 8x - 4$$
$$4 = 8x - 6x$$
$$4 = 2x$$
$$x = 2$$
$$x - 0.5 = 1.5$$

Sally will reach Rosalie in $1\frac{1}{2}$ hours or 90 minutes.

UNIT 7

1. $(3y)^2 = 3^2 y^2 = 9y^2$

2. $3x^0 = 3 \cdot 1 = 3$

3. $x^2(x^3)^4 = x^2 \cdot x^{12} = x^{2+12} = x^{14}$

4. $(x^2 y^3 z)^4 = x^8 y^{12} z^4$

5. $\left(\dfrac{x^2}{wz}\right)^3 = \dfrac{x^6}{w^3 z^3}$

6. $(2ab)b^2 = 2ab \cdot b^2 = 2 \cdot a \cdot b^{1+2} = 2ab^3$

7. $\dfrac{\left(x^2 y^3\right)^2}{5} = \dfrac{x^4 y^6}{5}$

8. $5(x^2 z)^2 = 5 \cdot x^4 z^2 = 5x^4 z^2$

9. $(5x^2 z)^2 = 5^2 x^4 z^2 = 25x^4 z^2$

10. $\left(\dfrac{a^2 b^3 c d^5}{3x^2 w^0}\right)^7 = \dfrac{a^{14} b^{21} c^7 d^{35}}{3^7 x^{14} w^0} = \dfrac{a^{14} b^{21} c^7 d^{35}}{2187 x^{14}}$

11. $\dfrac{\left(2ab\right)^2}{\left(3x^3\right)^2} = \dfrac{2^2 a^2 b^2}{3^2 x^6} = \dfrac{4a^2 b^2}{9x^6}$

12. $(3x^5)^2 (2x^3)^3 = 3^2 x^{10} \cdot 2^3 x^9 = 9x^{10} \cdot 8x^9 = 72x^{19}$

13. $(x^2 y)(xy^2) = x^{2+1} \cdot y^{1+2} = x^3 y^3$

14. $2(3ab^2)^2 = 2 \cdot 3^2 a^2 b^4 = 18a^2 b^4$

15. $(-4c)^2 = (-4)^2 c^2 = 16c^2$

16. $\left(\dfrac{xyz^2}{5a}\right)^3 = \dfrac{x^3 y^3 z^6}{5^3 a^3} = \dfrac{x^3 y^3 z^6}{125 a^3}$

17. $(-2abc)(bcd)(3abc^2) = -2 \cdot 3a^{1+1} \cdot b^{1+1+1} \cdot c^{1+1+2} \cdot d = -6a^2 b^3 c^4 d$

18. $(2x^2 yz)(-5xz)^2 (xyz^2)^3 = 2x^2 yz \cdot (-5)^2 x^2 z^2 \cdot x^3 y^3 z^6$
$= 2 \cdot 25 x^{2+2+3} \cdot y^{1+3} \cdot z^{1+2+6}$
$= 50x^7 y^4 z^9$

UNIT 8

1. $\dfrac{a^{-3}}{a^2} = \dfrac{1}{a^3 a^2} = \dfrac{1}{a^5}$

2. $\dfrac{a^{-2} x^3}{y^{-1}} = \dfrac{x^3 y}{a^2}$

3. $\left(x^2 y\right)^{-2} = x^{-4} y^{-2} = \dfrac{1}{x^4 y^2}$

4. $x^5 \cdot x^0 \cdot z^{-7} = \dfrac{x^5 \cdot 1}{z^7} = \dfrac{x^5}{z^7}$

5. $\dfrac{x^{-2}}{y^{-3}} = \dfrac{y^3}{x^2}$

6. $-2x^{-6}y^0 = \dfrac{-2y^0}{x^6} = \dfrac{-2(1)}{x^6} = \dfrac{-2}{x^6}$

7. $\left(3x^{-6}y^5\right)^{-2} = 3^{-2}x^{12}y^{-10} = \dfrac{x^{12}}{3^2 y^{10}} = \dfrac{x^{12}}{9y^{10}}$

8. $\dfrac{\left(ab^2\right)^{-3}}{\left(x^2 y^{-3}\right)^4} = \dfrac{a^{-3}b^{-6}}{x^8 y^{-12}} = \dfrac{y^{12}}{a^3 b^6 x^8}$

9. $\dfrac{\left(3ab^5\right)^{-3}}{2x^{-5}} = \dfrac{3^{-3}a^{-3}b^{-15}}{2x^{-5}} = \dfrac{x^5}{2\cdot 3^3 a^3 b^{15}} = \dfrac{x^5}{54 a^3 b^{15}}$

10. $\dfrac{7x^{-1}}{y^2} = \dfrac{7}{xy^2}$

11. $\left(5w^{-2}\right)^2\left(2w^{-2}\right) = 5^2 w^{-4}\cdot 2w^{-2} = \dfrac{5^2\cdot 2}{w^4\cdot w^2} = \dfrac{50}{w^6}$

12. $\dfrac{x^{-2}y^{-3}}{c^{-2}} = \dfrac{c^2}{x^2 y^3}$

13. $\dfrac{\left(5a^2 b^3\right)^2}{(-2x)^{-3}} = \dfrac{5^2 a^4 b^6}{(-2)^{-3}x^{-3}} = (-2)^3 5^2 a^4 b^6 x^3 = -200 a^4 b^6 x^3$

14. $\dfrac{16w^{-1}y^2 z^{-3}}{2x} = \dfrac{16y^2}{2wxz^3} = \dfrac{8y^2}{wxz^3}$

15. $\left[\dfrac{b^2}{(a^2 b)^{-2}}\right]^{-1} = \left[\dfrac{b^2}{a^{-4}b^{-2}}\right]^{-1} = \dfrac{b^{-2}}{a^4 b^2} = \dfrac{1}{a^4 b^2\cdot b^2} = \dfrac{1}{a^4 b^4}$

UNIT 9

1. $\dfrac{5a^7 b^2}{ab^{10}} = \dfrac{5a^{7-1}}{b^{10-2}} = \dfrac{5a^6}{b^8}$

2. $w^5\cdot w^0\cdot w^{-7} = \dfrac{w^5\cdot 1}{w^7} = \dfrac{w^5}{w^7} = \dfrac{1}{w^{7-5}} = \dfrac{1}{w^2}$

3. $\left(3a^4 b^{-2}\right)\left(a^5 b^{-3}\right) = \left(\dfrac{3a^4}{b^2}\right)\left(\dfrac{a^5}{b^3}\right) = \dfrac{3a^{4+5}}{b^{2+3}} = \dfrac{3a^9}{b^5}$

4. $\left(4x^{-3}y^7\right)\left(-2x^2 y^2\right) = \dfrac{4y^7\cdot(-2)x^2 y^2}{x^3} = \dfrac{-8y^7 y^2}{x^{3-2}} = \dfrac{-8y^{7+2}}{x} = \dfrac{-8y^9}{x}$

5. $\dfrac{x^{-4}}{x^4} = \dfrac{1}{x^4 x^4} = \dfrac{1}{x^{4+4}} = \dfrac{1}{x^8}$

6. $\dfrac{15x^5y^3}{3x^2y^7} = \dfrac{15\;x^5\;y^3}{3\;x^2\;y^7} = \dfrac{5x^{5-2}}{y^{7-3}} = \dfrac{5x^3}{y^4}$

7. $\dfrac{x^5\cdot x^{-4}}{x^{-3}} = \dfrac{x^5\cdot x^3}{x^4} = \dfrac{x^{5+3}}{x^4} = \dfrac{x^8}{x^4} = x^{8-4} = x^4$

8. $x\left(3x^2y^{-3}\right)^2 = x\cdot 3^2 x^4 y^{-6} = \dfrac{9x\cdot x^4}{y^6} = \dfrac{9x^5}{y^6}$

9. $\left(2w^{-2}\right)^2\left(5w^{-2}\right) = \left(2^2 w^{-4}\right)\left(5w^{-2}\right) = \left(\dfrac{4}{w^4}\right)\left(\dfrac{5}{w^2}\right) = \dfrac{4\cdot 5}{w^{4+2}} = \dfrac{20}{w^6}$

10. $x\left(5xy^{-2}\right)^{-2} = x\cdot 5^{-2}x^{-2}y^{+4} = \dfrac{x\,y^4}{5^2\,x^2} = \dfrac{y^4}{25x^{2-1}} = \dfrac{y^4}{25x}$

11. $\dfrac{m^{-9}s^{-8}}{m^{-4}s^3} = \dfrac{m^4}{m^9 s^3 s^8} = \dfrac{1}{m^{9-4}s^{3+8}} = \dfrac{1}{m^5 s^{11}}$

12. $\dfrac{\left(3x^2y\right)^{-1}}{2xy^{-5}} = \dfrac{3^{-1}x^{-2}y^{-1}}{2xy^{-5}} = \dfrac{y^5}{2x\cdot 3x^2y} = \dfrac{y^4}{6x^3}$

13. $\dfrac{\left(3xy^{-2}\right)^{-3}}{x} = \dfrac{3^{-3}x^{-3}y^6}{x} = \dfrac{y^6}{3^3 x^3\cdot x} = \dfrac{y^6}{27x^{3+1}} = \dfrac{y^6}{27x^4}$

14. $\left[\dfrac{\left(3x^2y\right)^3}{3x^7y^9}\right]^2 = \left[\dfrac{3^3 x^6 y^3}{3x^7y^9}\right]^2 = \dfrac{3^6 x^{12}y^6}{3^2 x^{14}y^{18}} = \dfrac{3^{6-2}}{x^{14-12}y^{18-6}} = \dfrac{3^4}{x^2y^{12}} = \dfrac{81}{x^2y^{12}}$

15. $\left[\dfrac{(xy)^2}{x^{-1}}\right]^3 = \dfrac{(xy)^6}{x^{-3}} = \dfrac{x^6y^6}{x^{-3}} = x^6y^6x^3 = x^{6+3}y^6 = x^9y^6$

16. $\left[\dfrac{(ab)^{-1}}{\left(a^{-2}b^3\right)^3}\right]^{-1} = \left[\dfrac{a^{-1}b^{-1}}{a^{-6}b^9}\right]^{-1} = \dfrac{ab}{a^6 b^{-9}} = \dfrac{abb^9}{a^6} = \dfrac{b^{1+9}}{a^{6-1}} = \dfrac{b^{10}}{a^5}$

UNIT 10

1. $\dfrac{2}{11} + \dfrac{1}{11} = \dfrac{2+1}{11} = \dfrac{3}{11}$

2. $\dfrac{7}{10} - \dfrac{9}{10} = \dfrac{7-9}{10} = \dfrac{-2}{10} = \dfrac{-1}{5}$

3. $\dfrac{7}{9} + \dfrac{1}{5} = \dfrac{7\cdot 5 + 9\cdot 1}{9\cdot 5} = \dfrac{35+9}{45} = \dfrac{44}{45}$

4. $\dfrac{1}{x} + 5 = \dfrac{1}{x} + \dfrac{5}{1} = \dfrac{1+5x}{x}$

5. $3 - \dfrac{5}{w} = \dfrac{3}{1} - \dfrac{5}{w} = \dfrac{3w - 1\cdot 5}{1\cdot w} = \dfrac{3w-5}{w}$

6. $7 + \dfrac{2}{x} = \dfrac{7}{1} + \dfrac{2}{x} = \dfrac{7x+2}{x}$

7. $\dfrac{2}{9} - \dfrac{-1}{10} = \dfrac{2\cdot 10 - 9\cdot -1}{9\cdot 10} = \dfrac{20+9}{90} = \dfrac{29}{90}$

8. $\dfrac{11}{t} + \dfrac{7}{r} = \dfrac{11r + 7t}{rt}$

9. $\dfrac{1}{x} + \dfrac{1}{x} = \dfrac{1+1}{x} = \dfrac{2}{x}$

10. $\dfrac{10}{x+1} + \dfrac{3}{x+1} = \dfrac{10+3}{x+1} = \dfrac{13}{x+1}$

11. $\dfrac{5}{a} - \dfrac{4}{a} = \dfrac{5-4}{a} = \dfrac{1}{a}$

12. $\dfrac{-s}{9} + \dfrac{k}{10} = \dfrac{-10s+9k}{90}$

13. $\dfrac{x}{2} - \dfrac{x}{5} = \dfrac{5x-2x}{2 \cdot 5} = \dfrac{3x}{10}$

14. $\begin{aligned} \dfrac{x+1}{2} - \dfrac{3}{5} &= \dfrac{5(x+1)-6}{10} \\[2mm] &= \dfrac{5x+5-6}{10} \\[2mm] &= \dfrac{5x-1}{10} \end{aligned}$

15. $\begin{aligned} \dfrac{x-1}{3} + \dfrac{x+1}{2} &= \dfrac{2(x-1)+3(x+1)}{3\cdot 2} \\[2mm] &= \dfrac{2x-2+3x+3}{6} = \dfrac{5x+1}{6} \end{aligned}$

16. $\begin{aligned} \dfrac{x+2}{2} - \dfrac{x+3}{3} &= \dfrac{3(x+2)-2(x+3)}{6} \\[2mm] &= \dfrac{3x+6-2x-6}{6} \\[2mm] &= \dfrac{x}{6} \end{aligned}$

UNIT 11

1. $\dfrac{\cancel{3}}{\cancel{8}} \cdot \dfrac{\cancel{32}^{\,4}}{\cancel{15}_{\,5}} = \dfrac{4}{5}$

2. $\dfrac{5}{18} \div \dfrac{3}{14} = \dfrac{5}{\cancel{18}_{\,9}} \cdot \dfrac{\cancel{14}^{\,7}}{3} = \dfrac{35}{27}$

3. $\dfrac{\cancel{3}}{\cancel{4}} \cdot \dfrac{\cancel{-8}^{\,-2}}{\cancel{9}_{\,3}} = \dfrac{-2}{3}$

4. $\dfrac{9}{14} \div \dfrac{5}{21} = \dfrac{9}{\cancel{14}_{\,2}} \cdot \dfrac{\cancel{21}^{\,3}}{5} = \dfrac{27}{10}$

5. $\dfrac{-8}{9} \div \dfrac{12}{-7} = \dfrac{\cancel{-8}^{\,-2}}{9} \cdot \dfrac{-7}{\cancel{12}_{\,3}} = \dfrac{-2 \cdot -7}{9 \cdot 3} = \dfrac{14}{27}$

6. $\left(\dfrac{2}{x} \cdot \dfrac{x}{5} \right) \div w = \left(\dfrac{2}{\cancel{x}} \cdot \dfrac{\cancel{x}}{5} \right) \div \dfrac{w}{1} = \dfrac{2}{5} \cdot \dfrac{1}{w} = \dfrac{2 \cdot 1}{5 \cdot w} = \dfrac{2}{5w}$

7. $\dfrac{\tfrac{3}{10}}{\tfrac{1}{10}} = \dfrac{3}{10} \div \dfrac{1}{10} = \dfrac{3}{\cancel{10}} \cdot \dfrac{\cancel{10}}{1} = 3$

8. $\dfrac{3-\dfrac{2}{5}}{3+\dfrac{2}{5}} = \left(\dfrac{3}{1} - \dfrac{2}{5}\right) \div \left(\dfrac{3}{1} + \dfrac{2}{5}\right)$

$$= \dfrac{15-2}{5} \div \dfrac{15+2}{5}$$

$$= \dfrac{13}{5} \div \dfrac{17}{5}$$

$$= \dfrac{13}{\cancel{5}} \cdot \dfrac{\cancel{5}}{17}$$

$$= \dfrac{13}{17}$$

9. $\dfrac{1-\dfrac{1}{3}}{\dfrac{5}{6}} = \left(1 - \dfrac{1}{3}\right) \div \dfrac{5}{6}$

$$= \left(\dfrac{1}{1} - \dfrac{1}{3}\right) \div \dfrac{5}{6}$$

$$= \left(\dfrac{3-1}{3}\right) \div \dfrac{5}{6}$$

$$= \dfrac{2}{3} \div \dfrac{5}{6}$$

$$= \dfrac{2}{\cancel{3}} \cdot \dfrac{\cancel{6}^{2}}{5}$$

$$= \dfrac{4}{5}$$

10. $\dfrac{\dfrac{a}{2}-\dfrac{3}{5}}{2} = \left(\dfrac{a}{2} - \dfrac{3}{5}\right) \div 2$

$$= \dfrac{5a-6}{10} \div \dfrac{2}{1}$$

$$= \dfrac{5a-6}{10} \cdot \dfrac{1}{2}$$

$$= \dfrac{5a-6}{20}$$

11. $7 \div \left(\dfrac{x-1}{4}\right) = \dfrac{7}{1} \cdot \dfrac{4}{x-1} = \dfrac{28}{x-1}$

12. $\dfrac{\dfrac{2}{x}-5}{x} = \left(\dfrac{2}{x} - \dfrac{5}{1}\right) \div x$

$$= \dfrac{2-5x}{x} \cdot \dfrac{1}{x}$$

$$= \dfrac{2-5x}{x^2}$$

13.

$$\frac{9 - \frac{x}{4}}{\frac{1}{2}} = \left(\frac{9}{1} - \frac{x}{4}\right) \div \frac{1}{2}$$

$$= \frac{36 - x}{4} \div \frac{1}{2}$$

$$= \frac{36 - x}{\cancel{4}_2} \cdot \frac{\cancel{2}}{1}$$

$$= \frac{36 - x}{2}$$

14.

$$\frac{x + \frac{x+2}{2}}{\frac{x}{2}} = \left(\frac{x}{1} + \frac{x+2}{2}\right) \div \frac{x}{2}$$

$$= \frac{2x + x + 2}{2} \div \frac{x}{2}$$

$$= \frac{3x+2}{2} \cdot \frac{2}{x}$$

$$= \frac{3x+2}{\cancel{2}} \cdot \frac{\cancel{2}}{x}$$

$$= \frac{3x+2}{x}$$

15.

$$\frac{\frac{a}{b} + 2}{\frac{a}{b} + 1} = \left(\frac{a}{b} + \frac{2}{1}\right) \div \left(\frac{a}{b} + \frac{1}{1}\right)$$

$$= \frac{a + 2b}{b} \div \frac{a + b}{b}$$

$$= \frac{a + 2b}{\cancel{b}} \cdot \frac{\cancel{b}}{a + b}$$

$$= \frac{a + 2b}{a + b}$$

UNIT 12

1. $\sqrt{5} \cdot \sqrt{20} = \sqrt{5 \cdot 20} = \sqrt{100} = 10$

2. $\sqrt{75} = \sqrt{25 \cdot 3} = 5\sqrt{3}$

3. $\sqrt{2a} \cdot \sqrt{3b} = \sqrt{6ab}$

4. $\sqrt{0} = 0$

5. $\sqrt{3} \cdot \sqrt{6} = \sqrt{18} = \sqrt{9 \cdot 2} = 3\sqrt{2}$

6. $\sqrt{64t^2} = 8t$

7. $\sqrt{w^4} = w^2$

8. $\sqrt{45} = \sqrt{9 \cdot 5} = 3\sqrt{5}$

9. $\sqrt{\dfrac{25}{x^2}} = \dfrac{5}{x}$

10. $\dfrac{\sqrt{20}}{\sqrt{5}} = \sqrt{\dfrac{20}{5}} = \sqrt{4} = 2$

11. $11\sqrt{2} + 3\sqrt{2} = 14\sqrt{2}$

12. $\sqrt{3x} \cdot \sqrt{3x} = \sqrt{9x^2} = 3x$

13. $7\sqrt{40} - 2\sqrt{10} = 7\sqrt{4 \cdot 10} - 2\sqrt{10}$
$$= 7 \cdot 2\sqrt{10} - 2\sqrt{10}$$
$$= 14\sqrt{10} - 2\sqrt{10}$$
$$= 12\sqrt{10}$$

14. $\sqrt{12y^8} = \sqrt{4y^8 \cdot 3} = 2y^4\sqrt{3}$

15. $\sqrt{50x^4} = \sqrt{25x^4 \cdot 2} = 5x^2\sqrt{2}$

16. $\sqrt{3}(\sqrt{2}+1) = \sqrt{6} + \sqrt{3}$

17. $\sqrt{5}(\sqrt{5} + \sqrt{3}) = \sqrt{25} + \sqrt{15} = 5 + \sqrt{15}$

18. $\sqrt{2a^2c} \cdot \sqrt{2ac} = \sqrt{4a^3c^2} = \sqrt{4a^2c^2 \cdot a}$
$$= 2ac\sqrt{a}$$

19. $\dfrac{\sqrt{48}}{\sqrt{18}} = \sqrt{\dfrac{48}{18}} = \sqrt{\dfrac{8}{3}} = \dfrac{\sqrt{8}}{\sqrt{3}} \cdot \dfrac{\sqrt{3}}{\sqrt{3}} = \dfrac{\sqrt{24}}{3}$
$$= \dfrac{\sqrt{4 \cdot 6}}{3} = \dfrac{2\sqrt{6}}{3}$$

20. $\dfrac{\sqrt{21x}}{\sqrt{15x^3}} = \sqrt{\dfrac{21x}{15x^3}} = \sqrt{\dfrac{7}{5x^2}} = \dfrac{\sqrt{7}}{\sqrt{5x^2}}$
$$= \dfrac{\sqrt{7}}{x\sqrt{5}} \cdot \dfrac{\sqrt{5}}{\sqrt{5}} = \dfrac{\sqrt{35}}{5x}$$

UNIT 13

1. $125^{1/3} = \sqrt[3]{125} = 5$

2. $(-1)^{2/3} = \left(\sqrt[3]{-1}\right)^2 = (-1)^2 = 1$

3. $\sqrt{-4}$
   No solution; we cannot take the square root of a negative number in the set of reals.

4. $4^{3/2} = \left(\sqrt{4}\right)^3 = \left(2^3\right) = 8$

5. $4^{1/2} = \sqrt{4} = 2$

6. $4^{-1/2} = \left(\sqrt{4}\right)^{-1} = 2^{-1} = \dfrac{1}{2}$

7. $x^{-1/2} = \dfrac{1}{x^{1/2}} = \dfrac{1}{\sqrt{x}}$

8. $x^{1/3} = \sqrt[3]{x}$

9. $a^{2/5} = \sqrt[5]{a^2}$

10. $4^{-3/2} = \left(\sqrt{4}\right)^{-3} = 2^{-3} = \dfrac{1}{2^3} = \dfrac{1}{8}$

11. $(x + 1)^{1/2} = \sqrt{x + 1}$

12. $x^{8/3} = \sqrt[3]{x^8} = \sqrt[3]{x^3 \cdot x^3 \cdot x^2} = x \cdot x\sqrt[3]{x^2} = x^2\sqrt[3]{x^2}$

13. $(4x)^{1/2} = \sqrt{4x} = 2\sqrt{x}$

14. $x^{11/2} = \sqrt{x^{11}} = \sqrt{x^{10} \cdot x} = x^5\sqrt{x}$

15. $(5x)^{-1/2} = \dfrac{1}{(5x)^{1/2}} = \dfrac{1}{\sqrt{5x}}$

16. $(18x^3)^{1/2} = \sqrt{18x^3} = \sqrt{9 \cdot 2 \cdot x^2 x} = \sqrt{9x^2 \cdot 2x} = 3x\sqrt{2x}$

17. $(2x)^{2/3} = \sqrt[3]{(2x)^2} = \sqrt[3]{4x^2}$

18. $(-64)^{2/3} = \left(\sqrt[3]{-64}\right)^2 = (-4)^2 = 16$

19. $\sqrt{7x} = (7x)^{1/2}$

20. $\sqrt[3]{2x} = (2x)^{1/3}$

21. $3\sqrt[4]{x} = 3x^{1/4}$

22. $5\sqrt[3]{y^2} = 5y^{2/3}$

23. $\dfrac{4}{\sqrt{x}} = \dfrac{4}{x^{1/2}}$

24. $\dfrac{1}{7\sqrt[3]{z}} = \dfrac{1}{7z^{1/3}}$

C25. 1.872

C26. $\sqrt[4]{17} = (17)^{1/4} = 2.031$

Press $\boxed{1}\ \boxed{7}\ \boxed{\wedge}\ \boxed{(}\ \boxed{1}\ \boxed{\div}\ \boxed{4}\ \boxed{)}\ \boxed{\text{ENTER}}$.

UNIT 14

1. $\dfrac{y^{2/3}}{y^{1/3}} = y^{2/3 - 1/3} = y^{1/3} = \sqrt[3]{y}$

2. $\left(y^{3/5}\right)^{1/4} = y^{3/5 \cdot 1/4} = y^{3/20} = \sqrt[20]{y^3}$

3. $x^{1/2} \cdot x^{2/5} = x^{1/2 + 2/5} = x^{(5+4)/10} = x^{9/10} = \sqrt[10]{x^9}$

4. $\left(\dfrac{a^4}{c^2}\right)^{1/2} = \dfrac{a^2}{c}$

5. $\left[\left(\sqrt{4}\right)^{-1}\right]^2 = \left[(2)^{-1}\right]^2 = (2)^{-2} = \dfrac{1}{2^2} = \dfrac{1}{4}$

6. $\left(\sqrt{x}\right)^{1/2} = \left(x^{1/2}\right)^{1/2} = x^{1/2 \cdot 1/2} = x^{1/4} = \sqrt[4]{x}$

7. $(8x^2)^{1/3} = 8^{1/3} x^{2/3} = 2\sqrt[3]{x^2}$

8. $\left(\dfrac{2^{-3} \cdot 2^5}{2^{-2}}\right)^3 = \dfrac{2^{-9} \cdot 2^{15}}{2^{-6}}$

$\qquad = \dfrac{2^{15} \cdot 2^6}{2^9}$

$\qquad = \dfrac{2^{21}}{2^9}$

$\qquad = 2^{21-9}$

$\qquad = 2^{12}$

An alternative, shorter approach would be

$\left(\dfrac{2^{-3} \cdot 2^5}{2^{-2}}\right)^3 = \left(\dfrac{2^2}{2^{-2}}\right)^3$

$\qquad = \left(2^2 \cdot 2^2\right)^3$

$\qquad = \left(2^4\right)^3$

$\qquad = 2^{12}$

9. $\left(\dfrac{x^{1/3}}{x^{2/3}}\right)^3 = \dfrac{x}{x^2} = \dfrac{1}{x}$

10. $x^{1/2} \cdot x^{5/2} = x^{1/2+5/2} = x^{6/2} = x^3$

11. $\left(\sqrt[3]{x^2}\right)^{1/2} = \left(x^{2/3}\right)^{1/2} = x^{1/3} = \sqrt[3]{x}$

12. $\dfrac{x^{-7/2} \cdot x^{3/2}}{\sqrt{x} \cdot x^{-3/2}} = \dfrac{x^{-7/2} \cdot x^{3/2}}{x^{1/2} \cdot x^{-3/2}}$

$= \dfrac{x^{3/2} \cdot x^{3/2}}{x^{7/2} \cdot x^{1/2}}$

$= \dfrac{x^{3/2+3/2}}{x^{7/2+1/2}}$

$= \dfrac{x^{6/2}}{x^{8/2}}$

$= \dfrac{x^3}{x^4}$

$= \dfrac{1}{x}$

13. $\left(8\sqrt{x}\right)^{-2/3} = \left(8x^{1/2}\right)^{-2/3}$

$= 8^{-2/3}\,x^{-1/3}$

$= \dfrac{1}{8^{2/3}\,x^{1/3}}$

$= \dfrac{1}{\left(\sqrt[3]{8}\right)^2 \sqrt[3]{x}}$

$= \dfrac{1}{4\sqrt[3]{x}}$

14. $\left(\dfrac{27^{5/3} \cdot 27^{-1/3}}{27^{1/3}}\right)^2 = \dfrac{27^{10/3} \cdot 27^{-2/3}}{27^{2/3}}$

$= \dfrac{27^{10/3}}{27^{2/3} \cdot 27^{2/3}}$

$= \dfrac{27^{10/3}}{27^{2/3+2/3}}$

$= \dfrac{27^{10/3}}{27^{4/3}}$

$= 27^{10/3 - 4/3}$

$= 27^{6/3}$

$= 27^2$

15. $\left(\dfrac{3x^{-1}}{\sqrt{x}}\right)^2 = \left(\dfrac{3x^{-1}}{x^{1/2}}\right)^2$

$= \dfrac{3^2 x^{-2}}{x}$

$= \dfrac{9}{x \cdot x^2}$

$= \dfrac{9}{x^3}$

UNIT 15

1. $\left(x^2 + 1\right)^{-2} = \dfrac{1}{\left(x^2 + 1\right)^2}$

2. $x^{-1} + 2^{-2} = \dfrac{1}{x} + \dfrac{1}{2^2}$

$= \dfrac{1}{x} + \dfrac{1}{4}$

$= \dfrac{4+x}{4x}$

3. $x^{-1} - 1 = \dfrac{1}{x} - \dfrac{1}{1} = \dfrac{1-x}{x}$

4. $\dfrac{2}{x^{-1} + 3y^{-1}} = \dfrac{2}{\dfrac{1}{x} + \dfrac{3}{y}}$

$= 2 \div \left(\dfrac{1}{x} + \dfrac{3}{y}\right)$

$= 2 \div \left(\dfrac{y + 3x}{xy}\right)$

$= \dfrac{2}{1} \cdot \dfrac{xy}{y + 3x}$

$= \dfrac{2xy}{y + 3x}$

5. $3^{-1} + 3^{-2} = \dfrac{1}{3} + \dfrac{1}{3^2} = \dfrac{1}{3} + \dfrac{1}{9}$

$\qquad = \dfrac{9+3}{27} = \dfrac{\cancel{12}^{\,4}}{\cancel{27}_{\,9}} = \dfrac{4}{9}$

6. $x^{-1} + 2y^{-1} = \dfrac{1}{x} + \dfrac{2}{y}$ ← Be careful that the 2 is in the numerator.

$\qquad = \left(\dfrac{1}{x} + \dfrac{2}{y} \right)$

$\qquad = \dfrac{y + 2x}{xy}$

7. $\left(ab^{-1} \right)^{-2} = a^{-2}b^2 = \dfrac{b^2}{a^2}$

8. $5(x+y)^{-1} = \dfrac{5}{(x+y)}$

9. $\dfrac{x^{-1}}{(2x-3)^{-2}} = \dfrac{(2x-3)^2}{x}$

10. $3x^{-2} + y = \dfrac{3}{x^2} + y$

$\qquad = \dfrac{3}{x^2} + \dfrac{y}{1}$

$\qquad = \dfrac{3 + x^2 y}{x^2}$

11. $\dfrac{a - a^{-1}}{a + a^{-1}} = \dfrac{a - \dfrac{1}{a}}{a + \dfrac{1}{a}}$

$\qquad = \left(a - \dfrac{1}{a} \right) \div \left(a + \dfrac{1}{a} \right)$

$\qquad = \left(\dfrac{a}{1} - \dfrac{1}{a} \right) \div \left(\dfrac{a}{1} + \dfrac{1}{a} \right)$

$\qquad = \left(\dfrac{a^2 - 1}{a} \right) \div \left(\dfrac{a^2 + 1}{a} \right)$

$\qquad = \dfrac{a^2 - 1}{\cancel{a}} \cdot \dfrac{\cancel{a}}{a^2 + 1}$

$\qquad = \dfrac{a^2 - 1}{a^2 + 1}$

12. $\dfrac{3^{-1} + 2^{-1}}{3^{-1} - 2^{-1}} = \dfrac{\dfrac{1}{3} + \dfrac{1}{2}}{\dfrac{1}{3} - \dfrac{1}{2}}$

$\qquad = \left(\dfrac{1}{3} + \dfrac{1}{2} \right) \div \left(\dfrac{1}{3} - \dfrac{1}{2} \right)$

$\qquad = \dfrac{2+3}{6} \div \dfrac{2-3}{6}$

$\qquad = \dfrac{5}{6} \div \dfrac{-1}{6}$

$\qquad = \dfrac{5}{\cancel{6}} \cdot \dfrac{\cancel{6}}{-1}$

$\qquad = -5$

13. $\dfrac{a + b^{-1}}{ab} = \dfrac{a + \dfrac{1}{b}}{ab}$

$\qquad = \left(a + \dfrac{1}{b} \right) \div ab$

$\qquad = \left(\dfrac{a}{1} + \dfrac{1}{b} \right) \div \dfrac{ab}{1}$

$\qquad = \dfrac{ab + 1}{b} \div \dfrac{ab}{1}$

$\qquad = \dfrac{ab + 1}{b} \cdot \dfrac{1}{ab}$

$\qquad = \dfrac{ab + 1}{ab^2}$

14. $\dfrac{3ab}{a^{-1} + b} = \dfrac{3ab}{\dfrac{1}{a} + b}$

$\qquad = 3ab \div \left(\dfrac{1}{a} + b \right)$

$\qquad = 3ab \div \left(\dfrac{1}{a} + \dfrac{b}{1} \right)$

$\qquad = 3ab \div \left(\dfrac{1}{a} + \dfrac{b}{1} \right)$

$\qquad = 3ab \div \left(\dfrac{1 + ab}{a} \right)$

$\qquad = 3ab \cdot \left(\dfrac{a}{1 + ab} \right)$

$\qquad = \dfrac{3ab}{1} \cdot \dfrac{a}{1 + ab}$

$\qquad = \dfrac{3a^2 b}{1 + ab}$

UNIT 16

1. $2cx^2(5c^2 - c - 3x) = 10c^3x^2 - 2c^2x^2 - 6cx^3$

2. $(x + 4)(x + 5) = x^2 + 5x + 4x + 20 = x^2 + 9x + 20$

3. $(x - 7)(x - 2) = x^2 - 2x - 7x + 14 = x^2 - 9x + 14$

4. $(x - 1)(x - 5) = x^2 - 5x - 1x + 5 = x^2 - 6x + 5$

5. $(x + 2)(x - 3) = x^2 - 3x + 2x - 6 = x^2 - x - 6$

6. $(a + 5)^2 = a^2 + 2(5a) + 25$
    $$= a^2 + 10a + 25$$

7. $(x + 2)(x - 2) = x^2 - 2x + 2x - 4 = x^2 - 4$

8. $(x - 1)^2 = x^2 + 2(-1x) + 1 = x^2 - 2x + 1$

9. $(x - 4)(x + 3) = x^2 + 3x - 4x - 12 = x^2 - x - 12$

10. $(2x + 1)(x + 1) = 2x(x + 1) + 1(x + 1)$
    $$= 2x^2 + 2x + x + 1$$
    $$= 2x^2 + 3x + 1$$

11. $(2x - 5)(x + 4) = 2x(x + 4) - 5(x + 4)$
    $$= 2x^2 + 8x - 5x - 20$$
    $$= 2x^2 + 3x - 20$$

12. $(3x - 2)(x + 7) = 3x(x + 7) - 2(x + 7)$
    $$= 3x^2 + 21x - 2x - 14$$
    $$= 3x^2 + 19x - 14$$

13. $(5x + 1)(x + 2) = 5x(x + 2) + 1(x + 2)$
    $$= 5x^2 + 10x + x + 2$$
    $$= 5x^2 + 11x + 2$$

14. $(3x + 1)(3x - 1) = 3x(3x - 1) + 1(3x - 1)$
    $$= 9x^2 - 3x + 3x - 1$$
    $$= 9x^2 - 1$$

15. $(2x + 3)(4x + 1) = 2x(4x + 1) + 3(4x + 1)$
    $$= 8x^2 + 2x + 12x + 3$$
    $$= 8x^2 + 14x + 3$$

16. $(5x - 1)(x + 2) = 5x(x + 2) - 1(x + 2)$
    $$= 5x^2 + 10x - x - 2$$
    $$= 5x^2 + 9x - 2$$

17. $(2x + 3)^2 = (2x)^2 + 2(2x)(3) + (3)^2$
    $$= 4x^2 + 12x + 9$$

18. $(4x - 3)^2 = 16x^2 + 2(4x)(-3) + 9$
    $$= 16x^2 - 24x + 9$$

19. $(3a + b)(2a - b) = 3a(2a - b) + b(2a - b)$
$$= 6a^2 - 3ab + 2ab - b^2$$
$$= 6a^2 - ab - b^2$$

20. $(x + y)(x - y) = x(x - y) + y(x - y)$
$$= x^2 - xy + xy - y^2$$
$$= x^2 - y^2$$

21. $(3x + 1)(2x - 5) = 3x(2x - 5) + 1(2x - 5)$
$$= 6x^2 - 15x + 2x - 5$$
$$= 6x^2 - 13x - 5$$

22. $x(x - 4)^2 = x(x^2 - 8x + 16)$
$$= x^3 - 8x^2 + 16x$$

23. $3x^2 (2x + 5)^2 = 3x^2 \left((2x)^2 + 2(2x)(5) + 5^2 \right)$
$$= 3x^2 \left(4x^2 + 20x + 25 \right)$$
$$= 12x^4 + 60x^3 + 75x^2$$

24. $(x - 2)(x^3 - 4x^2 + 7x - 1) = x(x^3 - 4x^2 + 7x - 1) - 2(x^3 - 4x^2 + 7x - 1)$
$$= x^4 - 4x^3 + 7x^2 - x - 2x^3 + 8x^2 - 14x + 2$$
$$= x^4 - 6x^3 + 15x^2 - 15x + 2$$

25. $\left(x^2 + 1 \right)\left(x^2 - 3 \right) = x^2 \left(x^2 - 3 \right) + 1(x^2 - 3)$
$$= x^4 - 3x^2 + x^2 - 3$$
$$= x^4 - 2x^2 - 3$$

26. $(x + 2y)(x - 3y) = x(x - 3y) + 2y(x - 3y)$
$$= x^2 - 3xy + 2xy - 6y^2$$
$$= x^2 - xy - 6y^2$$

27. $(2a - 1)(3 - a) = 2a(3 - a) - 1(3 - a)$
$$= 6a - 2a^2 - 3 + a$$
$$= -2a^2 + 7a - 3$$

28. $(x^2 - 3x + 1)(x^3 - 2x) = x^2(x^3 - 2x) - 3x(x^3 - 2x) + 1(x^3 - 2x)$
$$= x^5 - 2x^3 - 3x^4 + 6x^2 + x^3 - 2x$$
$$= x^5 - 3x^4 - x^3 + 6x^2 - 2x$$

29. $\left(x^2 + 5 \right)(x - 3) = x^2 (x - 3) + 5(x - 3)$
$$= x^3 - 3x^2 + 5x - 15$$

30. $(5a - 3b)(-2a + 6b) = 5a(-2a + 6b) - 3b(-2a + 6b)$
$$= -10a^2 + 30ab + 6ab - 18b^2$$
$$= -10a^2 + 36ab - 18b^2$$

UNIT 17

1.
$$\begin{array}{r} x + 3 \\ x + 2 \overline{\smash{)}\; x^2 + 5x + 6} \\ \underline{-x^2 - 2x} \\ 3x + 6 \\ \underline{-3x - 6} \\ 0 \end{array}$$

2.
$$\begin{array}{r} 12 + \dfrac{49}{x - 4} \\ x - 4 \overline{\smash{)}\; 12x + 1} \\ \underline{-12x + 48} \\ 49 \end{array}$$

3.
$$\begin{array}{r} x + 3 + \dfrac{1}{x + 5} \\ x + 5 \overline{\smash{)}\; x^2 + 8x + 16} \\ \underline{-x^2 - 5x} \\ 3x + 16 \\ \underline{-3x - 15} \\ 1 \end{array}$$

4.
$$\begin{array}{r} 2x - 3 + \dfrac{4}{5x + 1} \\ 5x + 1 \overline{\smash{)}\; 10x^2 - 13x + 1} \\ \underline{-10x^2 - 2x} \\ -15x + 1 \\ \underline{+15x + 3} \\ 4 \end{array}$$

5.
$$\begin{array}{r} x - 4 \\ 3x + 2 \overline{\smash{)}\; 3x^2 - 10x - 8} \\ \underline{-3x^2 - 2x} \\ -12x - 8 \\ \underline{+12x + 8} \\ 0 \end{array}$$

6.
$$\begin{array}{r} 3x - 8 + \dfrac{2}{2x - 1} \\ 2x - 1 \overline{\smash{)}\; 6x^2 - 19x + 10} \\ \underline{-6x^2 + 3x} \\ -16x + 10 \\ \underline{+16x - 8} \\ 2 \end{array}$$

7.
$$\begin{array}{r} 2x + 7 \\ 2x - 7 \overline{\smash{)}\; 4x^2 + 0x - 49} \\ \underline{-4x^2 + 14x} \\ 14x - 49 \\ \underline{-14x + 49} \\ 0 \end{array}$$

8.
$$\begin{array}{r} 5x + 5 + \dfrac{4}{x - 1} \\ x - 1 \overline{\smash{)}\; 5x^2 + 0x - 1} \\ \underline{-5x^2 + 5x} \\ +5x - 1 \\ \underline{-5x + 5} \\ 4 \end{array}$$

9.
$$\begin{array}{r} x^2 - x + 1 \\ x + 1 \overline{\smash{)}\; x^3 + 0x^2 + 0x + 1} \\ \underline{-x^3 - 1x^2} \\ -x^2 + 0x + 1 \\ \underline{x^2 + 1x} \\ x + 1 \\ \underline{-x - 1} \\ 0 \end{array}$$

10.
$$\begin{array}{r} x^2 - 2x + 4 + \dfrac{-6}{x + 2} \\ x + 2 \overline{\smash{)}\; x^3 + 0x^2 + 0x + 2} \\ \underline{-x^3 - 2x^2} \\ -2x^2 + 0x \\ \underline{+2x^2 + 4x} \\ 4x + 2 \\ \underline{-4x - 8} \\ -6 \end{array}$$

UNIT 18

1. $(x + 3)(x - 1)$

2. $(x - 8)(x - 7)$

3. $x(x + 1)$

4. $3xy(x - 4y)$

5. $3b(x^2 + 9b)$

6. $(x - 3)(x + 2)$

7. $(x + 3)(x + 2)$

8. $(x - 3)(x - 4)$

9. Prime

10. $(x + 2)(x - 4)$

11. Prime

12. $2x(x^2 + x + 11)$

13. $5(x^2 - x - 1)$

14. $(x + 5)(x - 2)$

15. $(x + 10)(x - 3)$

16. $(x + 6)(x + 1)$

17. $(x + 1)(x + 1)$

18. $(x - 3)(x - 3)$

19. $(x - 8)(x + 7)$

20. $(x + 9)(x - 5)$

21. $(x + 8)(x + 8)$

22. $(x - 5)(x - 8)$

23. $(x + 9)(x - 2)$

24. $x(x^2 + x + 5)$

25. $(x - 7)(x - 3)$

26. $(x - 9)(x + 2)$

27. $7x^2 - 14x + 7 = 7(x^2 - 2x + 1)$
$$= 7(x - 1)(x - 1)$$

28. $2x^3 - 6x^2 - 36x = 2x(x^2 - 3x - 18)$
$$= 2x(x - 6)(x + 3)$$

UNIT 19

1. $x^2 - 64 = (x + 8)(x - 8)$

2. $w^2 - 81 = (w + 9)(w - 9)$

3. $x^2 - y^2 = (x + y)(x - y)$

4. $16x^2 - 9 = (4x + 3)(4x - 3)$

5. $3b^2 - 75 = 3(b^2 - 25) = 3(b + 5)(b - 5)$

6. $x^2 - 6x + 9 = (x - 3)(x - 3)$

7. $2x^2 - 2 = 2(x^2 - 1) = 2(x + 1)(x - 1)$

8. $3abc^2 - 3abd^2 = 3ab(c^2 - d^2)$
$$= 3ab(c + d)(c - d)$$

9. $x^2 + 2x - 8 = (x + 4)(x - 2)$

10. $x^2 - x + 7$ prime

11. $x^3 - 36x = x(x^2 - 36) = x(x + 6)(x - 6)$

12. Prime, not the difference of two squares

13. $x^2 + 13x + 30 = (x + 10)(x + 3)$

14. $3r^3 - 6r^2 - 45r = 3r(r^2 - 2r - 15)$
    $$= 3r(r - 5)(r + 3)$$

15. $x^2 + 5x - 14 = (x + 7)(x - 2)$

16. $2a^2b^2c^2 - 4ab^2c^2 + 2b^2c^2 = 2b^2c^2(a^2 - 2a + 1)$
    $$= 2b^2c^2(a - 1)(a - 1)$$

17. $5x^2y - 15xy - 10y = 5y(x^2 - 3x - 2)$

18. $5x^4 + 10x^3 - 15x^2 = 5x^2(x^2 + 2x - 3)$
    $$= 5x^2(x + 3)(x - 1)$$

19. $3x^2 - 12 = 3(x^2 - 4) = 3(x - 2)(x + 2)$

20. $a^2b^2 - a^2c^2 = a^2(b^2 - c^2)$
    $$= a^2(b + c)(b - c)$$

21. $2xy^2 - 54xy + 100x = 2x(y^2 - 27y + 50)$
    $$= 2x(y - 2)(y - 25)$$

22. $10ab^2 - 140ab + 330a = 10a(b^2 - 14b + 33)$
    $$= 10a(b - 11)(b - 3)$$

23. $w^2x^2y^2 + 7w^2x^2y - 18w^2x^2 = w^2x^2(y^2 + 7y - 18)$
    $$= w^2x^2(y + 9)(y - 2)$$

24. $2ax^2 - 2ax - 40a = 2a(x^2 - x - 20)$
    $$= 2a(x - 5)(x + 4)$$

25. $4a^2 - 9b^2 = (2a + 3b)(2a - 3b)$

26. $2x^2 - 10x - 12 = 2(x^2 - 5x - 6)$
    $$= 2(x - 6)(x + 1)$$

27. $4r^3s^2 - 48r^2s^2 + 108rs^2 = 4rs^2(r^2 - 12r + 27)$
    $$= 4rs^2(r - 3)(r - 9)$$

28. $2y^2z + 38yz + 96z = 2z(y^2 + 19y + 48)$
    $$= 2z(y + 3)(y + 16)$$

29. $3a^2b^5 - 3a^2b = 3a^2b(b^4 - 1) = 3a^2b(b^2 + 1)(b^2 - 1)$
    $$= 3a^2b(b^2 + 1)(b + 1)(b - 1)$$

30. $a^2x^4 - 81a^2 = a^2(x^4 - 81)$
    $$= a^2(x^2 + 9)(x^2 - 9)$$
    $$= a^2(x^2 + 9)(x + 3)(x - 3)$$

UNIT 20

1. $7x^2 + 10x + 3 = 7x^2 + 7x + 3x + 3$

   | product = 21 |
   | sum   = 10 |
   | 7 and 3 |

   $\qquad\qquad = 7x(x + 1) + 3(x + 1)$
   $\qquad\qquad = (x + 1)(7x + 3)$

2. $2y^2 + 5y - 3 = 2y^2 + 6y - y - 3$

   | product = −6 |
   | sum   = 5 |
   | 6 and −1 |

   $\qquad\qquad = 2y(y + 3) - 1(y + 3)$
   $\qquad\qquad = (y + 3)(2y - 1)$

3. $6x^2 + 11x + 4 = 6x^2 + 3x + 8x + 4$

   | product = 24 |
   | sum   = 11 |
   | 3 and 8 |

   $\qquad\qquad = 3x(2x + 1) + 4(2x + 1)$
   $\qquad\qquad = (2x + 1)(3x + 4)$

4. $3x^3 - 5x^2 - 9x + 15 = x^2(3x - 5) - 3(3x - 5)$
   $\qquad\qquad\qquad\qquad = (3x - 5)(x^2 - 3)$

5. $4x^2 - 28x + 48 = 4(x^2 - 7x + 12)$
   $\qquad\qquad\qquad = 4(x - 3)(x - 4)$

6. $6x^2 + 13x + 6 = 6x^2 + 9x + 4x + 6$

   | product = 36 |
   | sum   = 13 |
   | 9 and 4 |

   $\qquad\qquad = 3x(2x + 3) + 2(2x + 3)$
   $\qquad\qquad = (2x + 3)(3x + 2)$

7. $2x^2 + 2x - 24 = 2(x^2 + x - 12)$
   $\qquad\qquad\qquad = 2(x + 4)(x - 3)$

8. $4x^3 - 10x^2 - 6x + 9 = 2x^2(2x - 5) - 3(2x - 3)$

   The technique has shown that the expression cannot be factored. The expression is prime.

9. $5x^2 - 4x - 1 = 5x^2 - 5x + 1x - 1$

$$\boxed{\begin{array}{l} \text{product} = -5 \\ \text{sum} \qquad = -4 \\ -5 \text{ and } 1 \end{array}} \begin{array}{l} = 5x(x-1) + 1(x-1) \\ = (x-1)(5x+1) \end{array}$$

10. $4x^2 + 8x + 4 = 4(x^2 + 2x + 1)$
$$= 4(x+1)(x+1)$$

11. $7x^2 + 13x - 2 = 7x^2 + 14x - 1x - 2$

$$\boxed{\begin{array}{l} \text{product} = -14 \\ \text{sum} \qquad = 13 \\ 14 \text{ and } -1 \end{array}} \begin{array}{l} = 7x(x+2) - 1(x+2) \\ = (x+2)(7x-1) \end{array}$$

12. $2x^2 - 7x + 6 = 2x^2 - 4x - 3x + 6$

$$\boxed{\begin{array}{l} \text{product} = 12 \\ \text{sum} \qquad = -7 \\ -4 \text{ and } -3 \end{array}} \begin{array}{l} = 2x(x-2) - 3(x-2) \\ = (x-2)(2x-3) \end{array}$$

13. $2y^2 - 17y + 35 = 2y^2 - 10y - 7y + 35$

$$\boxed{\begin{array}{l} \text{product} = 70 \\ \text{sum} \qquad = -17 \\ -10 \text{ and } -7 \end{array}} \begin{array}{l} = 2y(y-5) - 7(y-5) \\ = (y-5)(2y-7) \end{array}$$

14. $7x^2 + 32x - 15 = 7x^2 + 35x - 3x - 15$

$$\boxed{\begin{array}{l} \text{product} = -105 \\ \text{sum} \qquad = 32 \\ 35 \text{ and } -3 \end{array}} \begin{array}{l} = 7x(x+5) - 3(x+5) \\ = (x+5)(7x-3) \end{array}$$

15. $27x^2z - 3z = 3z(9x^2 - 1)$
$$= 3z(3x+1)(3x-1)$$

16. $6z^2 + 2z - 4 = 2(3z^2 + z - 2)$

$$\boxed{\begin{array}{l} \text{product} = -6 \\ \text{sum} \qquad = 1 \\ 3 \text{ and } -2 \end{array}} \begin{array}{l} = 2[3z^2 + 3z - 2z - 2] \\ = 2[3z(z+1) - 2(z+1)] \\ = 2[(z+1)(3z-2)] \\ = 2(z+1)(3z-2) \end{array}$$

17. $x^2z - 16xz + 64z = z(x^2 - 16x + 64)$
$$= z(x - 8)(x - 8)$$

18. $6xw^2 + 16wx - 6x = 2x(3w^2 + 8w - 3)$

product $= -9$
sum $\quad = 8$
9 and -1

$$= 2x(3w^2 + 9w - w - 3)$$
$$= 2x[3w(w + 3) - 1(w + 3)]$$
$$= 2x[(w + 3)(3w - 1)]$$
$$= 2x(w + 3)(3w - 1)$$

19. $8y + 4x + 2xy + x^2 = 4(2y + x) + x(2y + x)$
$$= (2y + x)(4 + x)$$

20. $2x^2 + 5x - 2 \qquad$ prime

product $\quad = -4$
sum $\qquad = 5$

21. $8x^2 + 30x - 27 = 8x^2 + 36x - 6x - 27$

product $= -216$
sum $\qquad = 30$
36 and -6

$$= 4x(2x + 9) - 3(2x + 9)$$
$$= (2x + 9)(4x - 3)$$

22. $xy^3 + 2y^2 - xy - 2 = y^2(xy + 2) - 1(xy + 2)$
$$= (xy + 2)(y^2 - 1)$$
$$= (xy + 2)(y + 1)(y - 1)$$

23. $12x^2 - 4x - 5 = 12x^2 - 10x + 6x - 5$

product $\quad = -60$
sum $\qquad = -4$
-10 and 6

$$= 2x(6x - 5) + 1(6x - 5)$$
$$= (6x - 5)(2x + 1)$$

24. $x^4 - y^4 = (x^2 + y^2)(x^2 - y^2)$
$$= (x^2 + y^2)(x + y)(x - y)$$

25. $1 - a^4 = (1 + a^2)(1 - a^2)$
$$= (1 + a^2)(1 + a)(1 - a)$$

UNIT 21

1. $x^2 + 5x - 14 = 0$
$(x - 2)(x + 7) = 0$
$x - 2 = 0$ or $x + 7 = 0$
     $x = 2$ $x = -7$

2. $x^2 + 13x + 30 = 0$
$(x + 10)(x + 3) = 0$
$x + 10 = 0$ or $x + 3 = 0$
     $x = -10$ $x = -3$

3. $6x^2 + 26 = 80$
     $6x^2 = 54$
     $x^2 = 9$
     $x = \pm\sqrt{9}$
     $x = \pm 3$
$x = 3$ or $x = -3$

4. $4x^2 + 8x + 4 = 0$
$4(x^2 + 2x + 1) = 0$
$4(x + 1)(x + 1) = 0$
     $x + 1 = 0$
       $x = -1$

5. $x^2 + 5x = 0$
$x(x + 5) = 0$
$x = 0$ or $x + 5 = 0$
          $x = -5$

6. $x^2 + 2x = 8$
$x^2 + 2x - 8 = 0$
$(x + 4)(x - 2) = 0$
$x + 4 = 0$ or $x - 2 = 0$
     $x = -4$ $x = 2$

7. $x^3 + 5x^2 + 6x = 0$
$x(x^2 + 5x + 6) = 0$
$x(x + 3)(x + 2) = 0$
$x = 0$ or $x + 3 = 0$ or $x + 2 = 0$
           $x = -3$ $x = -2$

8. $5x^2 - 5x = 0$
$5x(x - 1) = 0$
$5x = 0$ or $x - 1 = 0$
   $x = 0$ $x = 1$

9. $2x^2 - 7x + 3 = 0$
$2x^2 - 6x - 1x + 3 = 0$
$2x(x - 3) - 1(x - 3) = 0$
$(x - 3)(2x - 1) = 0$
$x - 3 = 0$ or $2x - 1 = 0$
   $x = 3$ $2x = 1$
                 $x = \dfrac{1}{2}$

10. $2x^2 + 8x + 6 = 0$
$2(x^2 + 4x + 3) = 0$
$2(x + 1)(x + 3) = 0$
$x + 1 = 0$ or $x + 3 = 0$
     $x = -1$ $x = -3$

11. $z^2 + 4z - 21 = 0$
$(z - 3)(z + 7) = 0$
$z - 3 = 0$ or $z + 7 = 0$
   $z = 3$ $z = -7$

12. $10x - 10 = 19x - x^2$
$x^2 - 9x - 10 = 0$
$(x - 10)(x + 1) = 0$
$x - 10 = 0$ or $x + 1 = 0$
   $x = 10$ $x = -1$

13. $3x^2 + 2x = 0$
$x(3x + 2) = 0$
$x = 0$ or $3x + 2 = 0$
            $3x = -2$
              $x = \dfrac{-2}{3}$

14.
$$2 - 2x^2 = 0$$
$$2 = 2x^2$$
$$1 = x^2$$
$$\pm 1 = x$$
$$x = 1 \quad \text{or} \quad x = -1$$

15.
$$2w^2 + 7w - 4 = 0$$
$$2w^2 + 8w - 1w - 4 = 0$$
$$2w(w + 4) - 1(w + 4) = 0$$
$$(w + 4)(2w - 1) = 0$$

$$w + 4 = 0 \quad \text{or} \quad 2w - 1 = 0$$
$$w = -4 \qquad\qquad 2w = 1$$
$$w = \frac{1}{2}$$

16.
$$x^3 + 3x^2 - 10x = 0$$
$$x(x^2 + 3x - 10) = 0$$
$$x(x + 5)(x - 2) = 0$$
$$x = 0 \quad \text{or} \quad x + 5 = 0 \quad \text{or} \quad x - 2 = 0$$
$$x = 0 \qquad\qquad x = -5 \qquad\qquad x = 2$$

17.
$$(x + 1)(x - 7)(x - 3) = 0$$
$$x + 1 = 0 \quad \text{or} \quad x - 7 = 0 \quad \text{or} \quad x - 3 = 0$$
$$x = -1 \qquad\qquad x = 7 \qquad\qquad x = 3$$

18.
$$2x^3 - x^2 + 14x - 7 = 0$$
$$x^2(2x - 1) + 7(2x - 1) = 0$$
$$(2x - 1)(x^2 + 7) = 0$$
$$2x - 1 = 0 \quad \text{or} \quad x^2 + 7 = 0$$
$$2x = 1 \qquad\qquad \text{no solution to this equation}$$
$$x = \frac{1}{2}$$

19.
$$x^4 + 16x^3 + 64x^2 = 0$$
$$x^2(x^2 + 16x + 64) = 0$$
$$x^2(x + 8)(x + 8) = 0$$
$$x^2 = 0 \quad \text{or} \quad x + 8 = 0 \quad \text{or} \quad x + 8 = 0$$
$$x = 0 \qquad\qquad x = -8 \qquad\qquad x = -8$$

20.
$$12x^2 + 5x - 2 = 0$$
$$12x^2 + 8x - 3x - 2 = 0$$
$$4x(3x + 2) - 1(3x + 2) = 0$$
$$(3x + 2)(4x - 1) = 0$$

$$3x + 2 = 0 \quad \text{or} \quad 4x - 1 = 0$$
$$3x = -2 \qquad\qquad 4x = 1$$
$$x = \frac{-2}{3} \qquad\qquad x = \frac{1}{4}$$

21.
$$\frac{x^2}{3} + 5 = 2$$
$$\frac{x^2}{3} = -3$$
$$x^2 = -9$$

Since $-9 < 0$, no real solution.

22.
$$\frac{x}{5} = \frac{3x + 20}{x + 15}$$
$$x(x + 15) = 5(3x + 20)$$
$$x^2 + 15x = 15x + 100$$
$$x^2 = 100$$
$$x = \pm\sqrt{100}$$
$$x = \pm 10$$
$$x = 10 \quad \text{or} \quad x = -10$$

UNIT 22

1. $x^2 + 3x - 1 = 0$

   $a = 1, b = 3, c = -1$

   $x = \dfrac{-(3) \pm \sqrt{(3)^2 - 4(1)(-1)}}{2(1)}$

   $x = \dfrac{-3 \pm \sqrt{9 + 4}}{2}$

   $x = \dfrac{-3 \pm \sqrt{13}}{2}$

2. $2x^2 - 3x - 2 = 0$

   $a = 2, b = -3, c = -2$

   $x = \dfrac{-(-3) \pm \sqrt{(-3)^2 - 4(2)(-2)}}{2(2)}$

   $x = \dfrac{3 \pm \sqrt{9 + 16}}{4}$

   $x = \dfrac{3 \pm \sqrt{25}}{4}$

   $x = \dfrac{3 \pm 5}{4}$

   $x = \dfrac{3 + 5}{4}$ or $x = \dfrac{3 - 5}{4}$

   $x = \dfrac{8}{4}$ $\qquad x = \dfrac{-2}{4}$

   $x = 2$ $\qquad x = \dfrac{-1}{2}$

3. $4x^2 + 8x - 8 = 0$

   $x^2 + 2x - 2 = 0$ (Divided by 4.)

   $a = 1, b = 2, c = -2$

   $x = \dfrac{-(2) \pm \sqrt{(2)^2 - 4(1)(-2)}}{2(1)}$

   $x = \dfrac{-2 \pm \sqrt{4 + 8}}{2}$

   $x = \dfrac{-2 \pm \sqrt{12}}{2}$

   $x = \dfrac{-2 \pm 2\sqrt{3}}{2}$

   $x = \dfrac{-2}{2} \pm \dfrac{2\sqrt{3}}{2}$

   $x = -1 \pm \sqrt{3}$

4. $2x^2 - 3x - 1 = 0$

Thus $a = 2$, $b = -3$, $c = -1$.

$$x = \frac{-(-3) \pm \sqrt{(-3)^2 - 4(2)(-1)}}{2(2)}$$

$$x = \frac{3 \pm \sqrt{9 + 8}}{4}$$

$$x = \frac{3 \pm \sqrt{17}}{4}$$

5. $6x^2 - 13x - 5 = 0$

$a = 6$, $b = -13$, $c = -5$

$$x = \frac{-(-13) \pm \sqrt{(-13)^2 - 4(6)(-5)}}{2(6)}$$

$$x = \frac{13 \pm \sqrt{169 + 120}}{12}$$

$$x = \frac{13 \pm \sqrt{289}}{12}$$

$$x = \frac{13 \pm 17}{12}$$

$$x = \frac{13 + 17}{12} \qquad \text{or} \qquad x = \frac{13 - 17}{12}$$

$$x = \frac{30}{12} \qquad\qquad\qquad x = \frac{-4}{12}$$

$$x = \frac{5}{2} \qquad\qquad\qquad x = \frac{-1}{3}$$

6. $5\left(\dfrac{1}{5}x^2 - 5x + 1 = 0\right)$

$x^2 - 25x + 5 = 0$

Thus $a = 1$, $b = -25$, $c = 5$.

$$x = \frac{-(-25) \pm \sqrt{(-25)^2 - 4(1)(5)}}{2(1)}$$

$$x = \frac{25 \pm \sqrt{625 - 20}}{2}$$

$$x = \frac{25 \pm \sqrt{605}}{2}$$

$$x = \frac{25 \pm \sqrt{121 \cdot 5}}{2}$$

$$x = \frac{25 \pm 11\sqrt{5}}{2}$$

7. $10x^2 + 13x = 3$

$10x^2 + 13x - 3 = 0$

$a = 10,\ b = 13,\ c = -3$

$$x = \frac{-(13) \pm \sqrt{(13)^2 - 4(10)(-3)}}{2(10)}$$

$$x = \frac{-13 \pm \sqrt{169 + 120}}{20}$$

$$x = \frac{-13 \pm \sqrt{289}}{20}$$

$$x = \frac{-13 \pm 17}{20}$$

$$x = \frac{-13 + 17}{20} \qquad \text{or} \qquad x = \frac{-13 - 17}{20}$$

$$x = \frac{4}{20} \qquad\qquad\qquad x = \frac{-30}{20}$$

$$x = \frac{1}{5} \qquad\qquad\qquad x = \frac{-3}{2}$$

8. $x^2 - 2x + 2 = 0$

Thus, $a = 1,\ b = -2,\ c = 2$.

$$x = \frac{-(-2) \pm \sqrt{(-2)^2 - 4(1)(2)}}{2(1)}$$

$$x = \frac{2 \pm \sqrt{4 - 8}}{2}$$

$$x = \frac{2 \pm \sqrt{-4}}{2}$$

There is no real solution. We cannot take the square root of a negative number in the reals.

9. $2x^2 - x = 0$

$a = 2$, $b = -1$, $c = 0$

$$x = \frac{-(-1) \pm \sqrt{(-1)^2 - 4(2)(0)}}{2(2)}$$

$$x = \frac{1 \pm \sqrt{1 + 0}}{4}$$

$$x = \frac{1 \pm \sqrt{1}}{4}$$

$$x = \frac{1 \pm 1}{4}$$

$$x = \frac{1 + 1}{4} \qquad \text{or} \qquad x = \frac{1 - 1}{4}$$

$$x = \frac{2}{4} \qquad\qquad\qquad x = \frac{0}{4}$$

$$x = \frac{1}{2} \qquad\qquad\qquad x = 0$$

10. $9x^2 - 12x + 4 = 0$

$a = 9$, $b = -12$, $c = 4$

$$x = \frac{-(-12) \pm \sqrt{(-12)^2 - 4(9)(4)}}{2(9)}$$

$$x = \frac{12 \pm \sqrt{144 - 144}}{18}$$

$$x = \frac{12 \pm \sqrt{0}}{18}$$

$$x = \frac{12 \pm 0}{18}$$

$$x = \frac{12 + 0}{18} \qquad \text{or} \qquad x = \frac{12 - 0}{18}$$

$$x = \frac{12}{18} \qquad\qquad\qquad x = \frac{12}{18}$$

$$x = \frac{2}{3} \qquad\qquad\qquad x = \frac{2}{3}$$

11. $2x^2 + 7x + 9 = 0$

$a = 2$, $b = 7$, $c = 9$

$$x = \frac{-(7) \pm \sqrt{(7)^2 - 4(2)(9)}}{2(2)}$$

$$x = \frac{-7 \pm \sqrt{49 - 72}}{4}$$

$$x = \frac{-7 \pm \sqrt{-23}}{4}$$

There is no real solution.

12. $x^2 - 2x - 10 = 0$

Thus, $a = 1$, $b = -2$, $c = -10$.

$$x = \frac{-(-2) \pm \sqrt{(-2)^2 - 4(1)(-10)}}{2(1)}$$

$$x = \frac{2 \pm \sqrt{44}}{2}$$

$$x = \frac{2 \pm 2\sqrt{11}}{2} \qquad \text{since } \sqrt{44} = \sqrt{4 \cdot 11} = 2\sqrt{11}$$

$$x = \frac{\cancel{2}\left(1 \pm \sqrt{11}\right)}{\cancel{2}}$$

$$x = 1 \pm \sqrt{11}$$

13. $3x^2 + x - 3 = 0$

$a = 3$, $b = 1$, $c = -3$

$$x = \frac{-(1) \pm \sqrt{(1)^2 - 4(3)(-3)}}{2(3)}$$

$$x = \frac{-1 \pm \sqrt{1 + 36}}{6}$$

$$x = \frac{-1 \pm \sqrt{37}}{6}$$

14. $2x^2 + 3x - 4 = 0$

$a = 2$, $b = 3$, $c = -4$

$$x = \frac{-(3) \pm \sqrt{(3)^2 - 4(2)(-4)}}{2(2)}$$

$$x = \frac{-3 \pm \sqrt{9 + 32}}{4}$$

$$x = \frac{-3 \pm \sqrt{41}}{4}$$

15. $4x^5 + 11x^4 - 3x^3 = 0$

$x^3\left(4x^2 + 11x - 3\right) = 0$

$x^3 = 0 \qquad \text{or} \qquad 4x^2 + 11x - 3 = 0$

$\quad x = 0 \qquad\qquad a = 4$, $b = 11$, $c = -3$

$$x = \frac{-(11) \pm \sqrt{(11)^2 - 4(4)(-3)}}{2(4)}$$

$$x = \frac{-11 \pm \sqrt{121 + 48}}{8}$$

$$x = \frac{-11 \pm \sqrt{169}}{8}$$

$$x = \frac{-11 \pm 13}{8}$$

$$x = \frac{-11 + 13}{8} \quad \text{or} \quad x = \frac{-11 - 13}{8}$$

$$x = \frac{2}{8} \qquad\qquad x = \frac{-24}{8}$$

$$x = 0 \qquad x = \frac{1}{4} \qquad\qquad x = -3$$

16. $2x^4 + 2x^3 + 2x^2 = 0$

$2x^2(x^2 + x + 1) = 0$

$2x^2 = 0 \quad \text{or} \quad x^2 + x + 1 = 0$

$x^2 = 0 \qquad$ Use the quadratic formula with $a = 1, b = 1, c = 1$.

$$x = 0 \qquad\qquad x = \frac{-1 \pm \sqrt{1 - 4}}{2}$$

$$x = \frac{-1 \pm \sqrt{-3}}{2}$$

There is no real solution to this part since we cannot have a negative number under the square root symbol in the reals.

Therefore the only solution to this equation within the set of real numbers is $x = 0$.

UNIT 23

1. $x + y = 8$

$y = -x + 8$

x	y	(x, y)
0	8	(0, 8)
2	6	(2, 6)
4	4	(4, 4)
6	2	(6, 2)
8	0	(8, 0)

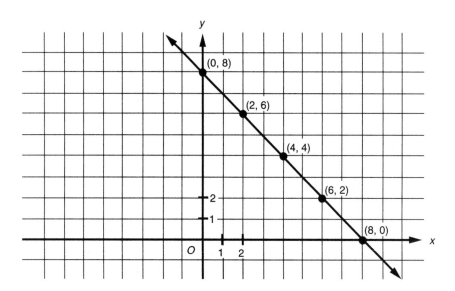

The x-intercept is 8.
The y-intercept is 8.
The slope is −1.

2.

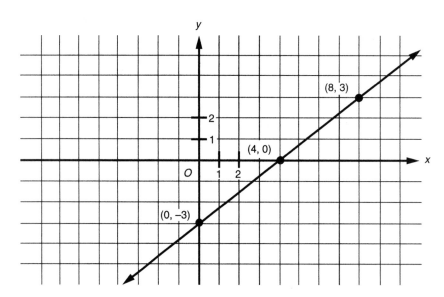

$3x - 4y = 12$

The x-intercept is where $y = 0$: $3x - 4(0) = 12$
$$3x = 12$$
$$x = 4$$

Therefore the x-intercept is 4.

The y-intercept is where $x = 0$: $3(0) - 4y = 12$
$$-4y = 12$$
$$y = -3$$

Therefore the y-intercept is −3.

From the graph, the slope is $\dfrac{3}{4}$.

3. $7x + y = 10$
   $y = -7x + 10$

   The y-intercept is 10. The slope is $\dfrac{-7}{1}$. Plot a point on the y-axis at 10. Move down 7 blocks and to the right 1 block; mark a point. Repeat.

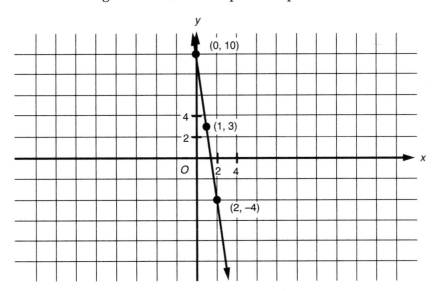

   The x-intercept is $\dfrac{10}{7}$.
   The y-intercept is 10.
   The slope is -7.

4. $y = 5$ is a horizontal line.

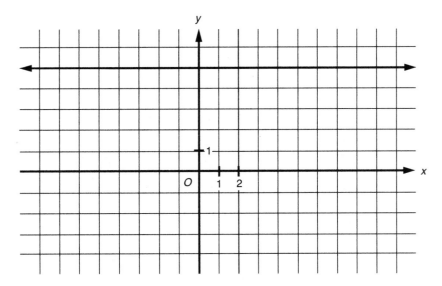

   There is no x-intercept.
   The y-intercept is 5.
   The slope is 0.

5. $y - 2x = 0$

$y = 2x + 0$

The y-intercept is 0. The slope is $\dfrac{2}{1}$. Plot a point on the y-axis at 0 (the origin). Move up 2 blocks and to the right 1 block; mark a point. Repeat.

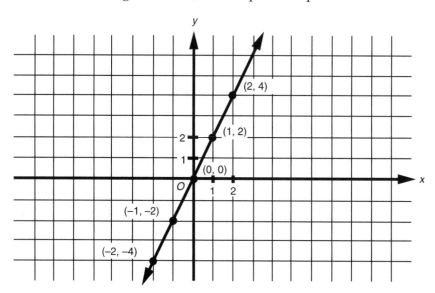

The x-intercept is 0.
The y-intercept is 0.
The slope is 2.

6. $x + 3y = 21$

$y = \dfrac{-1}{3}x + 7$

The y-intercept is 7. The slope is $\dfrac{-1}{3}$. Plot a point on the y-axis at 7. Move down 1 block and to the right 3 blocks; mark a point. Repeat.

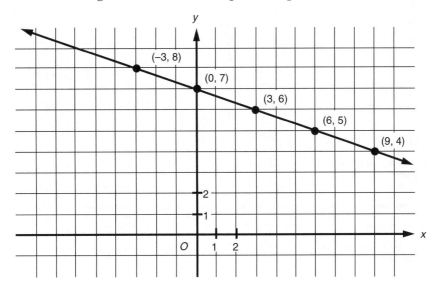

$x + 3y = 21$

$x + 3(0) = 21$

$x = 21;$ the x-intercept is 21.

$0 + 3y = 21$

$y = 7;$ the y-intercept is 7.

The slope is $\dfrac{-1}{3}$.

7. $y = \dfrac{2}{7}x + 4$

The y-intercept is 4. The slope is $\dfrac{2}{7}$. Plot a point on the y-axis at 4. Move up 2 blocks and to the right 7 blocks; mark a point. From the y-intercept, move down 2 blocks and to the left 7 blocks.

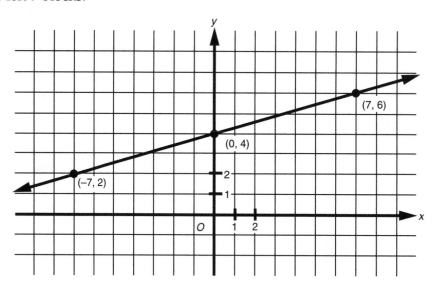

The x-intercept is −14.
The y-intercept is 4.
The slope is $\dfrac{2}{7}$.

8. $x = -3$ is a vertical line.

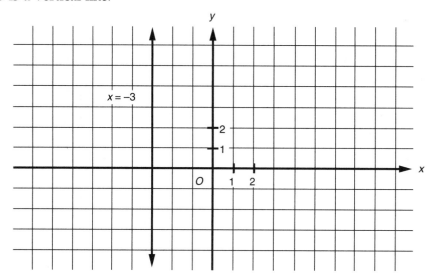

The x-intercept is −3.
There is no y-intercept.
The slope is undefined.

9.　　　$-3x = 2y + 4$

　　$-3x - 4 = 2y$

　$\dfrac{-3}{2}x - 2 = y$

x	y	(x, y)
-4	4	$(-4, 4)$
-2	1	$(-2, 1)$
0	-2	$(0, -2)$
2	-5	$(2, -5)$

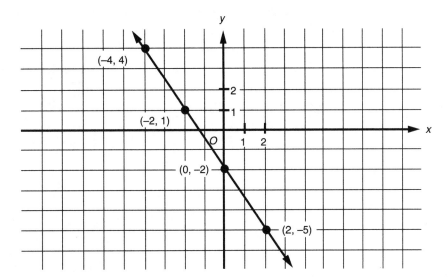

The x-intercept is $\dfrac{-4}{3}$.

The y-intercept is -2.

The slope is $\dfrac{-3}{2}$.

10. $y = \dfrac{2}{5}x - 1$

The y-intercept is –1. The slope is $\dfrac{2}{5}$. Plot a point on the y-axis at –1. Move up 2 blocks and to the right 5 blocks; mark a point. From the y-intercept, move down 2 blocks and to the left 5 blocks.

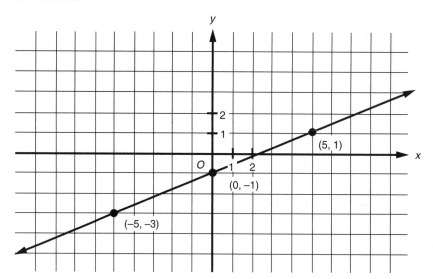

$$y = \frac{2}{5}x - 1$$

$$0 = \frac{2}{5}x - 1$$

$$0 = 2x - 5$$

$$-2x = -5$$

$$x = \frac{5}{2}$$

The x-intercept is $\dfrac{5}{2}$ or 2.5.

$$y = \frac{2}{5}(0) - 1$$

$$y = -1$$

The y-intercept is –1.

The slope is $\dfrac{2}{5}$.

C11. $y = \dfrac{3}{4}x - 5$

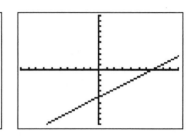

C12. $y = 1.23x + 5.76$

a.   Press $\boxed{\text{Y=}}$ $\boxed{\text{CLEAR}}$ $\boxed{1}$ $\boxed{.}$ $\boxed{2}$ $\boxed{3}$ $\boxed{\text{X,T,}\theta\text{,n}}$ $\boxed{+}$ $\boxed{5}$ $\boxed{.}$ $\boxed{7}$ $\boxed{6}$ $\boxed{\text{GRAPH}}$.

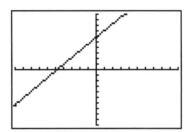

b.   To find the y-intercept, press $\boxed{\text{2nd}}$ $\boxed{\text{CALC}}$ $\boxed{1}$ $\boxed{0}$ $\boxed{\text{ENTER}}$.
     The y-intercept is $y = 5.76$.

c.   To find the x-intercept, press $\boxed{\text{2nd}}$ $\boxed{\text{CALC}}$ $\boxed{2}$.

At the <u>Left Bound?</u> prompt, move the cursor to the left of the

x-intercept and press $\boxed{\text{ENTER}}$.

At the <u>Right Bound?</u> prompt, move the cursor to the right of the

x-intercept and press $\boxed{\text{ENTER}}$.

At the <u>Guess?</u> prompt, move the cursor between the two boundaries

and press $\boxed{\text{ENTER}}$.
The x-intercept is –4.682927.

d.   To find y, press $\boxed{\text{2nd}}$ $\boxed{\text{CALC}}$ $\boxed{1}$ $\boxed{\text{(–)}}$ $\boxed{1}$ $\boxed{.}$ $\boxed{1}$ $\boxed{2}$ $\boxed{\text{ENTER}}$.
     The y-value is 4.3824.

UNIT 24

1. $y = x^2 + 6x + 8$ with $a = 1$, $b = 6$, and $c = 8$.

   The axis of symmetry is $x = \dfrac{-b}{2a} = \dfrac{-6}{2(1)} = -3$, or $x = -3$.

   Table of values

x	y
-6	8
-5	3
-4	0
-3	-1
-2	0
-1	3
0	8

   Answer:

   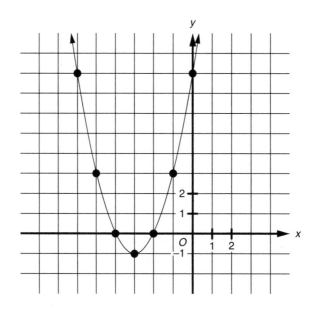

   Check:

   Is it $\smile$-shaped? ✓
   Does it open in the right direction? $a > 0$, so it should open up. ✓
   Does it pass through the y-intercept, $(0, 8)$? ✓

2. $y = x^2 + 2x - 8$ with $a = 1$, $b = 2$, $c = -8$.

The axis of symmetry is $x = \dfrac{-b}{2a} = \dfrac{-2}{2(1)} = -1$, or $x = -1$

Table of values

x	y
-4	0
-3	-5
-2	-8
-1	-9
0	-8
1	-5
2	0

Answer:

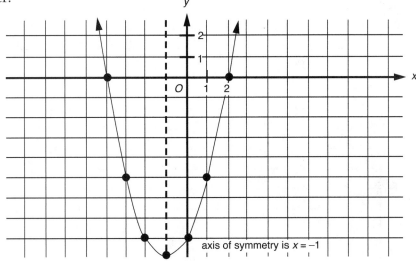

axis of symmetry is $x = -1$

Check:

Is it $\cup$-shaped?   ✓

Does it open in the right direction? $a > 0$, so it should open up.   ✓

Does it pass through the y-intercept, $(0, -8)$?   ✓

3. $y = \dfrac{-1}{2}x^2$ with $a = \dfrac{-1}{2}$, $b = 0$, and $c = 0$.

The axis of symmetry is $x = \dfrac{-b}{2a} = \dfrac{-0}{2\left(\dfrac{-1}{2}\right)} = 0$, or $x = 0$.

Table of values

x	y
-4	-8
-2	-2
0	0
2	-2
4	-8

Answer:

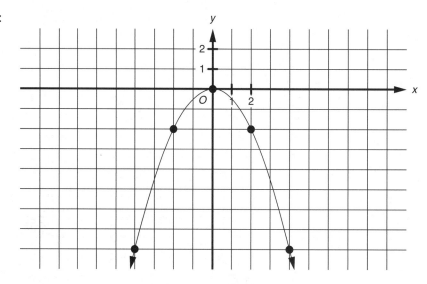

Check:

Is it $\smile$-shaped? ✓
Does it open in the right direction? $a < 0$, so it should open down. ✓
Does it pass through the y-intercept, $(0, 0)$? ✓

4. $y = -x^2 - 4x - 3$ with $a = -1$, $b = -4$, $c = -3$.

    The axis of symmetry is $x = \dfrac{-b}{2a} = \dfrac{-(-4)}{2(-1)} = -2$, or $x = -2$.

Table of values

x	y
–5	–8
–4	–3
–3	0
–2	1
–1	0
0	–3
1	–8

Answer:

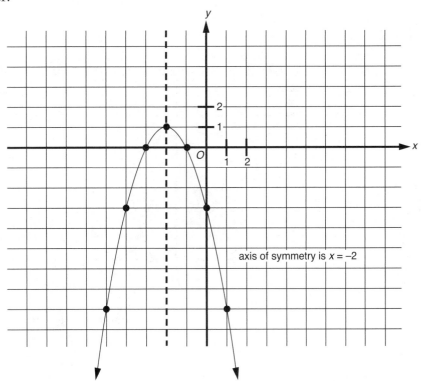

axis of symmetry is $x = -2$

Check:

Is it $\smile$-shaped? ✓
Does it open in the right direction? $a < 0$, so it should open down. ✓
Does it pass through the y-intercept, $(0, -3)$? ✓

5. $y = x^2 - 6x + 5$ with $a = 1$, $b = -6$, and $c = 5$.

The axis of symmetry is $x = \dfrac{-b}{2a} = \dfrac{-(-6)}{2(1)} = 3$, or $x = 3$.

Table of values

x	y
0	5
1	0
2	−3
3	−4
4	−3
5	0
6	5

Answer:

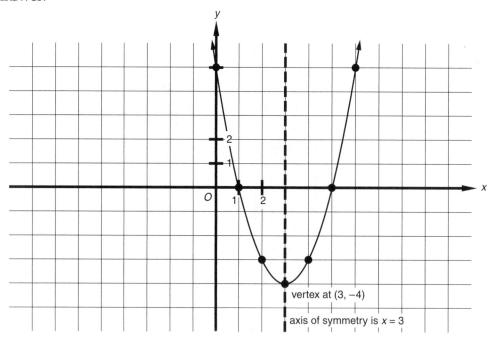

vertex at (3, −4)

axis of symmetry is $x = 3$

Check:

Is it ⌣-shaped? ✓
Does it open in the right direction? $a > 0$, so it should open up. ✓
Does it pass through the y-intercept, (0, 5)? ✓

6. $y = 4 + 2x^2$ with $a = 2$, $b = 0$, $c = 4$.

 The axis of symmetry is $x = \dfrac{-b}{2a} = \dfrac{-0}{2(2)} = 0$, or $x = 0$.

Table of values

x	y
-1	6
0	4
1	6

Answer:

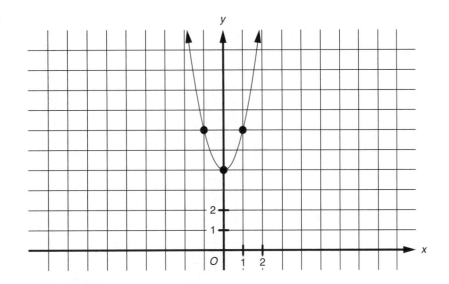

Check:

Is it $\smile$-shaped? ✓
Does it open in the right direction? $a > 0$, so it should open up. ✓
Does it pass through the y-intercept, $(0, 4)$? ✓

7. $y = x^2 - 4x + 4$ with $a = 1$, $b = -4$, $c = 4$.

The axis of symmetry is $x = \dfrac{-b}{2a} = \dfrac{-(-4)}{2(1)} = 2$, or $x = 2$.

Table of values

x	y
0	4
1	1
2	0
3	1
4	4

Answer:

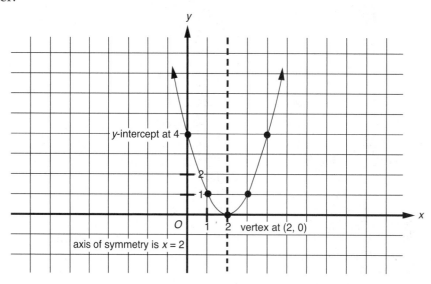

Check:

Is it $\smile$-shaped? ✓
Does it open in the right direction? $a > 0$, so it should open up. ✓
Does it pass through the y-intercept, $(0, 4)$? ✓

8. $y = 9 - x^2$ with $a = -1$, $b = 0$, $c = 9$.

The axis of symmetry is $x = \dfrac{-b}{2a} = \dfrac{-(0)}{2(-1)} = 0$, or $x = 0$.

Table of values

x	y
-3	0
-2	5
-1	8
0	9
1	8
2	5
3	0

Answer:

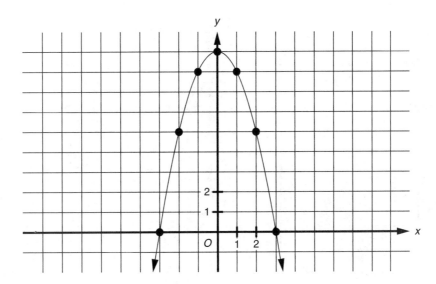

Check:

Is it $\smile$-shaped?   ✓
Does it open in the right direction? $a < 0$, so it should open down.   ✓
Does it pass through the y-intercept, $(0, 9)$?   ✓

9. $y = x^2 + 6x + 9$ with $a = 1$, $b = 6$, and $c = 9$.

The axis of symmetry is $x = \dfrac{-b}{2a} = \dfrac{-(6)}{2(1)} = -3$, or $x = -3$.

Table of values

x	y
-6	9
-5	4
-4	1
-3	0
-2	1
-1	4
0	9

Answer:

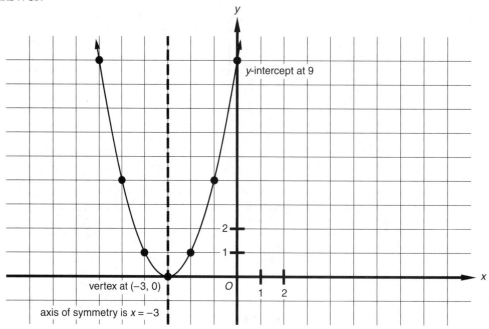

Check:

        Is it ⌣-shaped? ✓

        Does it open in the right direction? $a > 0$, so it should open up. ✓

        Does it pass through the y-intercept, $(0, 9)$? ✓

10. $y = 2x^2 + 4x - 1$ with $a = 2$, $b = 4$, $c = -1$.

The axis of symmetry is $x = \dfrac{-b}{2a} = \dfrac{-(4)}{2(2)} = -1$, or $x = -1$.

Table of values

x	y
-3	5
-2	-1
-1	-3
0	-1
1	5

Answer:

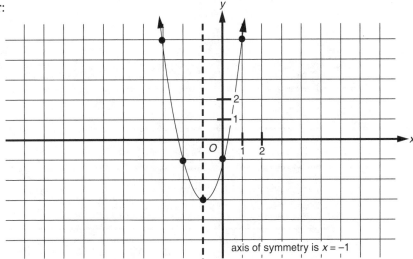

Check:

Is it $\smile$-shaped?   ✓
Does it open in the right direction? $a > 0$, so it should open up.   ✓
Does it pass through the y-intercept, $(0, -1)$?   ✓

11. $y = 5x^2 - 20x + 11$ with $a = 5$, $b = -20$, and $c = 11$.

The axis of symmetry is $x = \dfrac{-b}{2a} = \dfrac{-(-20)}{2(5)} = 2$, or $x = 2$.

Table of values

x	y
0	11
1	−4
2	−9
3	−4
4	11

Answer:

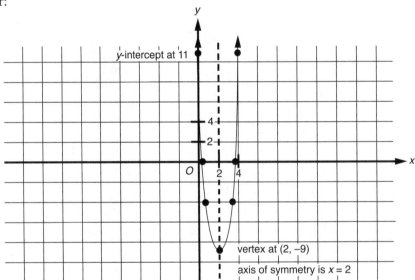

Check:
  Is it $\smile$-shaped?   ✓
  Does it open in the right direction? $a > 0$, so it should open up.   ✓
  Does it pass through the y-intercept, (0, 11)?   ✓

12. $y = -x^2 + 6x$ with $a = -1$, $b = 6$, $c = 0$.

The axis of symmetry is $x = \dfrac{-b}{2a} = \dfrac{-(6)}{2(-1)} = 3$, or $x = 3$.

Table of values

x	y
0	0
1	5
2	8
3	9
4	8
5	5
6	0

Answer:

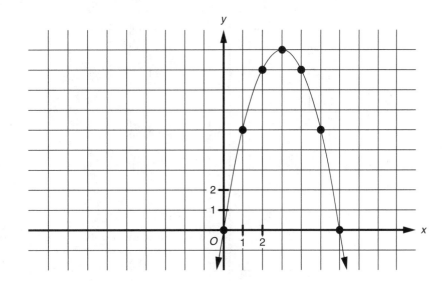

Check:
    Is it ⌣-shaped? ✓
    Does it open in the right direction? $a < 0$, so it should open down. ✓
    Does it pass through the y-intercept, $(0, 0)$? ✓

13. $y = 3x^2 - 3x + 2$ with $a = 3$, $b = -3$, and $c = 2$.

The axis of symmetry is $x = \dfrac{-b}{2a} = \dfrac{-(-3)}{2(3)} = \dfrac{1}{2} = 0.5$, or $x = 0.5$.

Table of values

x	y
−1	8
0	2
0.5	1.25
1	2
2	8

Answer:

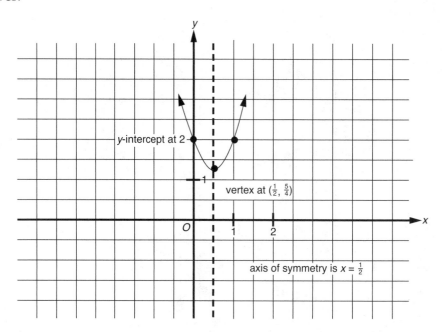

Check:

    Is it $\smile$-shaped? ✓

    Does it open in the right direction? $a > 0$, so it should open up. ✓

    Does it pass through the y-intercept, $(0, 2)$? ✓

14. $y = 2x^2 - 12x + 3$ with $a = 2$, $b = -12$, $c = 3$

The axis of symmetry is $x = \dfrac{-b}{2a} = \dfrac{-(-12)}{2(2)} = 3$, or $x = 3$.

Table of values

x	y
0	3
1	−7
2	−13
3	−15
4	−13
5	−7
6	3

Answer:

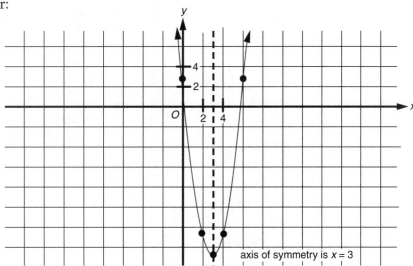

axis of symmetry is $x = 3$

Check:

Is it $\smile$-shaped? ✓
Does it open in the right direction? $a > 0$, so it should open up. ✓
Does it pass through the y-intercept, $(0, 3)$? ✓

15. $y = \dfrac{-1}{4}x^2 + 3x - 8$ with $a = \dfrac{-1}{4}$, $b = 3$, and $c = -8$.

The axis of symmetry is $x = \dfrac{-b}{2a} = \dfrac{-(3)}{2\left(\dfrac{-1}{4}\right)} = 6$, or $x = 6$.

Table of values

x	y
0	-8
2	-3
4	0
6	1
8	0
10	-3
12	-8

Answer:

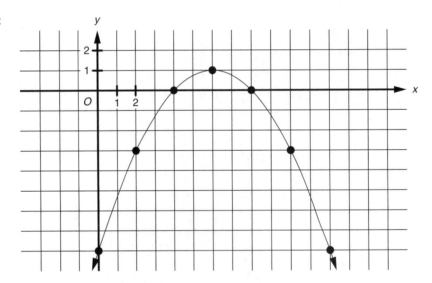

Check:

Is it $\smile$-shaped? ✓
Does it open in the right direction? $a < 0$, so it should open down. ✓
Does it pass through the y-intercept, $(0, -8)$? ✓

UNIT 25

Given: $f(x) = x - 2$

1. $f(7) = 7 - 2 = 5$

2. $f(-3) = -3 - 2 = -5$

3. $f(3.7) = 3.7 - 2 = 1.7$

4. $f(0) = 0 - 2 = -2$

Given: $g(x) = 3x$

5. $g(2) = 3(2) = 6$

6. $g(-5) = 3(-5) = -15$

7. $g(0) = 3(0) = 0$

8. $g(-4.1) = 3(-4.1) = -12.3$

8a. $g(x - 5) = 3(x - 5) = 3x - 15$

8b. $g(a^2) = 3(a^2) = 3a^2$

Given: $h(x) = 5x + 1$

9. $h(10) = 5(10) + 1 = 51$

10. $h(3) = 5(3) + 1 = 15 + 1 = 16$

11. $h(0) = 5(0) + 1 = 1$

12. $h(-4) = 5(-4) + 1 = -20 + 1 = -19$

12a. $h(k) = 5(k) + 1 = 5k + 1$

12b. $h(-x) = 5(-x) + 1 = -5x + 1$

12c. $h(x) = -14$
$$5x + 1 = -14$$
$$5x = -15$$
$$x = -3$$

Given: $f(x) = 1 - x^2$

13. $f(3) = 1 - (3)^2 = 1 - 9 = -8$

14. $f(-5) = 1 - (-5)^2 = 1 - 25 = -24$

15. $f(-1) = 1 - (-1)^2 = 1 - 1 = 0$

Given: $G = \{(5, 6), (7, 7), (8, 5), (6, 11)\}$

16. $D = \{5, 6, 7, 8\}$

17. $R = \{5, 6, 7, 11\}$

18. $G(8) = 5$

19. $G(5) = 6$

20. $G^{-1}(11) = 6$

21. $G^{-1}(5) = 8$

22. Yes

C23. $f(2.5) = 10.75$
$f(0.25) = 1.75$

C24. Given: $f(x) = 3x^4 + 2x^3 - x^2 + 1$

$f(2) = 61$
$f(1.5) = 20.6875$

UNIT 26

1. $\begin{cases} y = 3x + 2 \\ x - 3y = -6 \end{cases}$

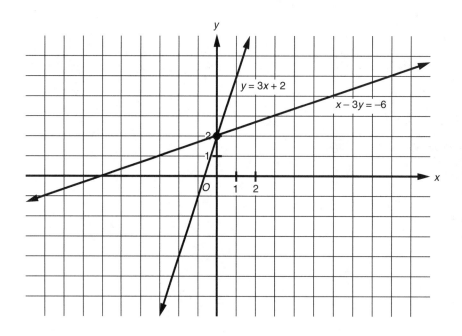

Solution is $(0, 2)$.

2. $\begin{cases} 2x + y = 8 \\ 2x - 3y = 0 \end{cases}$

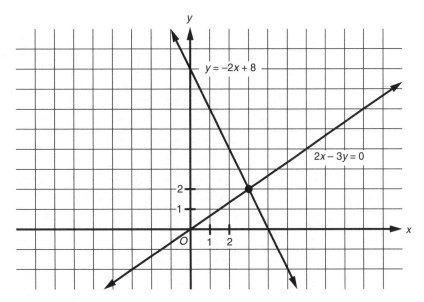

Solution is (3, 2).

3. $\begin{cases} y = 2x + 3 \\ y + 2 = 4x + 1 \end{cases}$

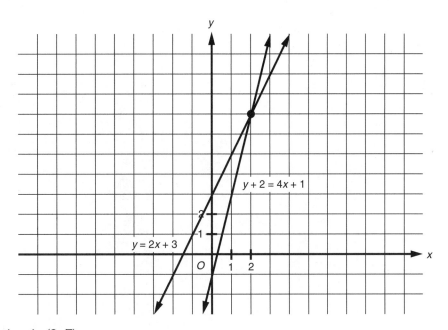

Solution is (2, 7).

4.
$$\begin{cases} y = -\dfrac{1}{2}x - 4 \\ x + 2y = 6 \end{cases}$$

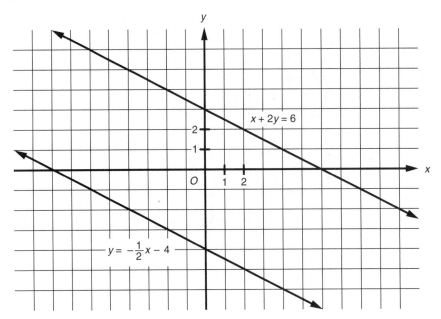

There is no solution.

5. $5x + 2y = 22$
    $3x - 2y = 10$
    $8x \qquad = 32$
    $\qquad x = 4$

If $x = 4$ and $5x + 2y = 22$,
$$5(4) + 2y = 22$$
$$20 + 2y = 22$$
$$2y = 2$$
$$y = 1$$

(4, 1)

6. $+2(5x + 3y = 1)$
    $5(-2x + 5y = 12)$

    $10x + 6y = 2$
    $\underline{-10x + 25y = 60}$
    $\qquad 31y = 62$
    $\qquad y = 2$

If $y = 2$ and $5x + 3y = 1$
$$5x + 3(2) = 1,$$
$$5x + 6 = 1,$$
$$5x = -5,$$
$$x = -1.$$

The solution to the system is $x = -1$ and $y = 2$.

7. $5(4x - 3y = -10)$
  $-4(5x + 2y = 22)$

  $20x - 15y = -50$
  $-20x - 8y = -88$
  $ -23y = -138$
  $y = 6$

  If $y = 6$ and $4x - 3y = -10$,
  $4x - 3(6) = -10$
  $4x - 18 = -10$
  $4x = 8$
  $x = 2$

  $(2, 6)$

8. $2x + y = 0$
  $\underline{x - y = 1}$
  $3x = 1$

  $x = \dfrac{1}{3}$

  If $x = \dfrac{1}{3}$ and $x - y = 1$,

  $$\frac{1}{3} - y = 1,$$

  $$-y = \frac{2}{3},$$

  $$y = -\frac{2}{3}.$$

  The solution to the system is $x = \dfrac{1}{3}$ and $y = -\dfrac{2}{3}$.

9. $11(-2x + 17y = 6)$
  $2(11x - 5y = -33)$
  $-22x + 187y = 66$
  $\underline{22x - 10y = -66}$
  $177y = 0$
  $y = 0$

  If $y = 0$ and $-2x + 17y = 6$,
  $-2x + 0 = 6$,
  $x = -3$.

  The solution to the system is $x = -3$ and $y = 0$.

10. $4(5x + 2y = 50)$
      $-5(4x - 3y = -52)$

$$20x + 8y = 200$$
$$-20x + 15y = 260$$
$$23y = 460$$
$$y = 20$$

If $y = 20$ and $5x + 2y = 50$,
$$5x + 2(20) = 50$$
$$5x + 40 = 50$$
$$5x = 10$$
$$x = 2$$

$(2, 20)$

11. $2\left(2x + \dfrac{1}{2}y = 2\right)$
      $6x - y = 1$

$$4x + y = 4$$
$$\underline{6x - y = 1}$$
$$10x \quad = 5$$
$$x = \dfrac{1}{2}$$

If $x = \dfrac{1}{2}$ and $6x - y = 1$,

$$6\left(\dfrac{1}{2}\right) - y = 1,$$
$$3 - y = 1,$$
$$2 = y.$$

The solution to the system is $x = \dfrac{1}{2}$ and $y = 2$.

12. $5\left(y = -\dfrac{2}{5}x + 4\right)$

$5y = -2x + 20$

$2x + 5y = 20$

$3\left(x = \dfrac{1}{3}y - 7\right)$

$3x = y - 21$

$3x - y = -21$

$2x + 5y = 20$

$5(3x - y = -21)$

$2x + 5y = 20$

$\underline{15x - 5y = -105}$

$17x = -85$

$x = -5$

If $x = -5$ and $y = -\dfrac{2}{5}x + 4$,

$$y = -\dfrac{2}{5}(-5) + 4,$$
$$= 2 + 4,$$
$$= 6.$$

The solution to the system is $x = -5$ and $y = 6$.

13. $6(3x - 2y = 8)$

$3(-6x + 4y = 10)$

$18x - 12y = 48$

$\underline{-18x + 12y = 30}$

$0 = 78$

There is no solution to the system.

14. $3(2a + 3b = 10)$

$-2(3a + 2b = 10)$

$6a + 9b = 30$

$\underline{-6a - 4b = -20}$

$5b = 10$

$b = 2$

If $b = 2$ and $2a + 3b = 10$,

$$2a + 3(2) = 10$$
$$2a + 6 = 10$$
$$2a = 4$$
$$a = 2$$

$a = 2$ and $b = 2$

15. $13x - 5y = -5$

$5(x + y = 1)$

$13x - 5y = -5$

$\underline{5x + 5y = 5}$

$18x \quad\quad = 0$

$x = 0$

If $x = 0$ and $x + y = 1$,

$0 + y = 1,$

$y = 1.$

The solution to the system is $x = 0$ and $y = 1$.

16. $-7x + y = 2$

$2x - 8 = y$

Since $y = 2x - 8$:

$-7x \quad\quad + y = 2$

$-7x + (2x - 8) = 2$

$-5x - 8 = 2$

$-5x = 10$

$x = -2$

If $x = -2$ and $y = 2x - 8$,

$y = 2(-2) - 8$

$y = -4 - 8$

$y = -12$

$(-2, -12)$

17. $x + 4y \quad\quad = 8$

$y = -\dfrac{1}{4}x - 7$

$x + 4y \quad\quad = 8$

$x + 4\left(-\dfrac{1}{4}x - 7\right) = 8$ by substitution

$x - x \quad\quad - 28 = 8$

$-28 = 8$

There is no solution to the system.

18. $x - 3y = 2$
       $4x - 10y = 10$

Since $x = 3y + 2$:

$4x \qquad - 10y = 10$
$4(3y + 2) - 10y = 10$
$\quad 12y + 8 - 10y = 10$
$\qquad\quad 2y + 8 = 10$
$\qquad\qquad 2y = 2$
$\qquad\qquad\ y = 1$

If $y = 1$ and $x = 3y + 2$,
$\qquad\qquad x = 3(1) + 2$
$\qquad\qquad x = 3 + 2$
$\qquad\qquad x = 5$

(5, 1)

19. $x = 7 - \dfrac{1}{2}y$
       $8x - y = -4$

Since $x = 7 - \dfrac{1}{2}y$:

$8x - \qquad\quad y = -4$
$8\left(7 - \dfrac{1}{2}y\right) - y = -4$
$\quad 56 - 4y - y = -4$
$\qquad 56 - 5y = -4$
$\qquad\quad -5y = -60$
$\qquad\qquad\ y = 12$

If $y = 12$ and $x = 7 - \dfrac{1}{2}y$,

$\qquad\qquad x = 7 - \dfrac{1}{2}(12)$
$\qquad\qquad x = 7 - 6$
$\qquad\qquad x = 1$

(1, 12)

20. $2x = -7(y + 1)$

Since $y = -\dfrac{2}{7}x - 1$:

$$2x = -7\left(-\dfrac{2}{7}x - 1 + 1\right)$$

$$2x = -7\left(-\dfrac{2}{7}x\right)$$

$$2x = 2x$$

$$0 = 0$$

There are infinitely many solutions to the system that satisfy $y = -\dfrac{2}{7}x - 1$.

UNIT 27

1. Use $x = 2$ to substitute:

$y = 3x^2 - 7x + 11$

$\quad = 3(2)^2 - 7(2) + 11$

$\quad = 12 - 14 + 11$

$\quad = 9$

The solution to the system is $x = 2$ and $y = 9$ or $(2, 9)$.

2. Use $y = 4x^2$ to substitute:

$-4x + y = 0$

$\downarrow$

$-4x + 4x^2 = 0$

$4x^2 - 4x = 0$

$4x(x - 1) = 0$

$4x = 0$	$x - 1 = 0$
$x = 0$	$x = 1$

If $\quad x = 0 \qquad$ If $\quad x = 1$

and $\quad y = 4x^2, \qquad$ and $\quad y = 4x^2,$

$\qquad y = 4(0)^2 \qquad\qquad\qquad y = 4(1)^2$

$\qquad\quad = 0. \qquad\qquad\qquad\qquad = 4.$

There are two solutions to the system.

$x = 0$ and $y = 0 \qquad\qquad (0, 0)$

and

$x = 1$ and $y = 4 \qquad\qquad (1, 4)$

3.
$$y = x^2 + 5x$$
$$y = -2x + 44$$
$$x^2 + 5x = -2x + 44$$
$$x^2 + 7x - 44 = 0$$
$$(x + 11)(x - 4) = 0$$

$x + 11 = 0$ or $x - 4 = 0$
$x = -11$ $x = 4$

$x = -11$	$x = 4$
$y = -2x + 44$	$y = -2x + 44$
$y = -2(-11) + 44$	$y = -2(4) + 44$
$y = 22 + 44$	$y = -8 + 44$
$y = 66$	$y = 36$

There are two solutions to the system:
$x = 4$ and $y = 36$ (4, 36)
and
$x = -11$ and $y = 66$ (−11, 66)

4.
$$y = x^2 + 5x - 21$$
$$y = x$$
$$x^2 + 5x - 21 = x$$
$$x^2 + 4x - 21 = 0$$
$$(x + 7)(x - 3) = 0$$

$x + 7 = 0$ or $x - 3 = 0$
$x = -7$ $x = 3$

$x = -7$	$x = 3$
$y = x$	$y = x$
$y = -7$	$y = 3$

There are two solutions to the system:
$x = 3$ and $y = 3$ (3, 3)
and
$x = -7$ and $y = -7$ (−7, −7)

5. Use $y = x^2 + 2x + 5$ to substitute:

    $$-4x + 2y = -7$$

    $$\downarrow$$

    $$-4x + 2(x^2 + 2x + 5) = -7$$
    $$-4x + 2x^2 + 4x + 10 = -7$$
    $$2x^2 = -17$$
    $$x^2 = \frac{-17}{2}$$

    There is no solution to the system.

6.

    $$y = 2x^2 - 4x$$
    $$y = 8x - 18$$
    $$2x^2 - 4x = 8x - 18$$
    $$2x^2 - 12x + 18 = 0$$
    $$2(x^2 - 6x + 9) = 0$$
    $$2(x - 3)(x - 3) = 0$$

$x - 3 = 0$	or	$x - 3 = 0$
$x = 3$		$x = 3$

$x = 3$	$x = 3$
$y = 8x - 18$	$y = 8x - 18$
$y = 8(3) - 18$	$y = 8(3) - 18$
$y = 24 - 18$	$y = 24 - 18$
$y = 6$	$y = 6$

    The solution to the system is $x = 3$ and $y = 6$ or (3, 6).

7. Use $y = 1$ to substitute:

    $$y = 3x^2 + 8x - 2$$
    $$\downarrow$$
    $$1 = 3x^2 + 8x - 2$$
    $$0 = 3x^2 + 8x - 3$$
    $$0 = (3x - 1)(x + 3)$$

$3x - 1 = 0$	$x + 3 = 0$
$3x = 1$	$x = -3$
$x = \dfrac{1}{3}$	

    There are two solutions to the system.

    $$x = \frac{1}{3} \text{ and } y = 1 \qquad \left(\frac{1}{3}, 1\right)$$

    and

    $$x = -3 \text{ and } y = 1 \qquad (-3, 1)$$

8.
$$y = x^2 + 13$$
$$y = -10x - 12$$
$$x^2 + 13 = -10x - 12$$
$$x^2 + 10x + 25 = 0$$
$$(x + 5)(x + 5) = 0$$

$$x + 5 = 0 \quad \text{or} \quad x + 5 = 0$$
$$x = -5 \qquad\qquad x = -5$$

$x = -5$	$x = -5$
$y = -10x - 12$	$y = -10x - 12$
$y = -10(-5) - 12$	$y = -10(-5) - 12$
$y = 50 - 12$	$y = 50 - 12$
$y = 38$	$y = 38$

The solution to the system is $x = -5$ and $y = 38$ or $(-5, 38)$.

9.　$y = -x^2 + 2x + 8$
　$x - y = -6$

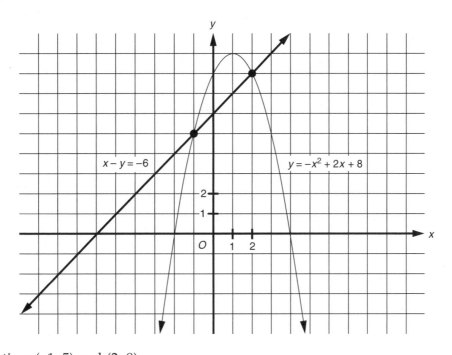

Solution: $(-1, 5)$ and $(2, 8)$

10. $y = x^2 - 4$

     $y + 2x = -1$

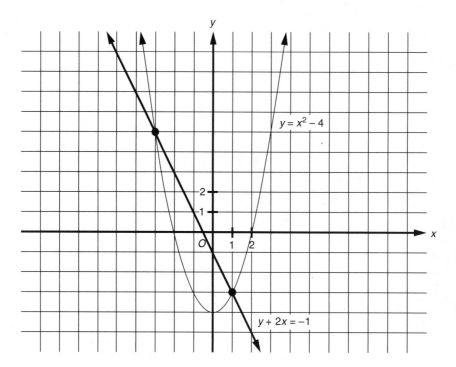

    Solution: $(-3, 5)$ and $(1, -3)$

UNIT 28

1. $10 + 2x > 12$

    $2x > 2$

     $x > 1$

2. $4 - (12 - 3x) \leq -5$

   $4 - 12 + 3x \leq -5$

    $-8 + 3x \leq -5$

      $3x \leq -5 + 8$

      $3x \leq 3$

       $x \leq 1$

3. $5x < 22 - (2x + 1)$

 $5x < 22 - 2x - 1$

 $5x < 21 - 2x$

 $7x < 21$

 $x < 3$

4. $4x + (3x - 7) > 2x - (28 - 2x)$

 $4x + 3x - 7 > 2x - 28 + 2x$

 $7x - 7 > 4x - 28$

 $3x > -21$

 $x > -7$

5. $5 - 3x \leq 23$

 $-3x \leq 18$

 $x \geq -6$

6. $3x + 4(x - 2) \geq x - 5 + 3(2x - 1)$

 $3x + 4x - 8 \geq x - 5 + 6x - 3$

 $7x - 8 \geq 7x - 8$

 $0 \geq 0$

The solution is the entire set of real numbers.

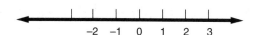

7. $3x - 2(5x + 2) > 1 - 5(x - 1) + x$

 $3x - 10x - 4 > 1 - 5x + 5 + x$

 $-7x - 4 > 6 - 4x$

 $-3x - 4 > 6$

 $-3x > 10$

 $x < \dfrac{-10}{3}$

8. $3x - 2(x - 5) < 3(x - 1) - 2x - 11$

    $3x - 2x + 10 < 3x - 3 - 2x - 11$

    $x + 10 < x - 14$

    $x - x < -14 - 10$

    $0 < -24$

    There is no solution.

9. $3x + 4(x - 2) + 7 > x - 5 + 3(2x - 1)$

    $3x + 4x - 8 + 7 > x - 5 + 6x - 3$

    $7x - 1 > 7x - 8$

    $-1 > -8$

    The solution is the entire set of real numbers.

10. $5x - 2(3x - 4) > 4[2x - 3(1 - 3x)]$

    $5x - 6x + 8 > 4[2x - 3 + 9x]$

    $-x + 8 > 4[11x - 3]$

    $-x + 8 > 44x - 12$

    $-x - 44x > -12 - 8$

    $-45x > -20$

    $x < \dfrac{20}{45} \text{ or } \dfrac{4}{9}$

    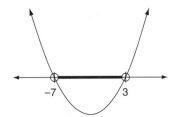

UNIT 29

1. $x^2 + 4x - 21 < 0$

    Let $y = x^2 + 4x - 21$ and graph. Since $a = 1$, the parabola opens up.

    $x^2 + 4x - 21 = 0$

    $(x - 3)(x + 7) = 0$

    $x = 3 \quad \text{or} \quad x = -7$

    The x-intercepts are -7 and 3.

    Make a rough sketch.

    The curve is below the x-axis ($y < 0$) for $-7 < x < 3$.

2. $x^2 + 2x - 8 < 0$

   Since $a = 1$, the parabola opens up.

   $x^2 + 2x - 8 = 0$

   $(x - 2)(x + 4) = 0$

   $x = 2$ or $x = -4$

   The x-intercepts are -4 and 2.

   The curve is below the x-axis ($y < 0$) for $-4 < x < 2$.

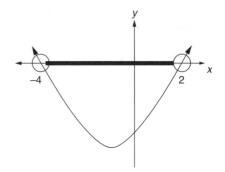

3. $x^2 - x - 12 > 0$

   Let $y = x^2 - x - 12$ and graph. Since $a = 1$, the parabola opens up.

   $x^2 - x - 12 = 0$

   $(x - 4)(x + 3) = 0$

   $x = 4$ or $x = -3$

   The x-intercepts are -3 and 4.

   Make a rough sketch.

   The curve is above the x-axis ($y > 0$) for $x < -3$ or $x > 4$.

4. $-x^2 - 4x - 3 < 0$

   Since $a = -1$, the parabola opens down.

   $-x^2 - 4x - 3 = 0$

   $x^2 + 4x + 3 = 0$

   $(x + 3)(x + 1) = 0$

   $x = -3$ or $x = -1$

   The x-intercepts are -3 and -1.

   The curve is below the x-axis ($y < 0$) for $x < -3$ or $x > -1$.

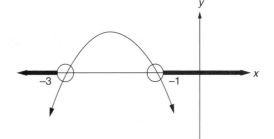

5. $-x^2 - x + 2 > 0$

   Let $y = -x^2 - x + 2$ and graph. Since $a = -1$, the parabola opens down.

   $-x^2 - x + 2 = 0$

   $x^2 + x - 2 = 0$

   $(x - 1)(x + 2) = 0$

   $x = 1$ or $x = -2$

   The x-intercepts are -2 and 1.

   Make a rough sketch.

   The curve is above the x-axis ($y > 0$) for $-2 < x < 1$.

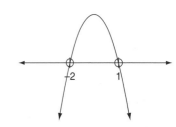

6. $-4x^2 \geq 0$

Since $a = -4$, the parabola opens down.

$$-4x^2 = 0$$
$$x^2 = 0$$
$$x = 0$$

The x-intercept is 0.

The curve is touching or above the x-axis ($y \geq 0$) only at $x = 0$.

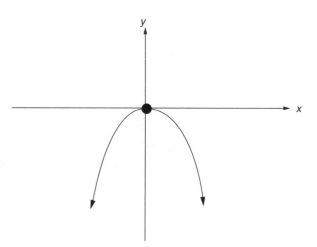

7. $x^2 - 4x + 4 \leq 0$

Let $y = x^2 - 4x + 4$ and graph. Since $a = 1$, the parabola opens up.

$$x^2 - 4x + 4 = 0$$
$$(x - 2)(x - 2) = 0$$
$$x = 2 \quad \text{or} \quad x = 2$$

The x-intercept is 2.

Make a rough sketch.

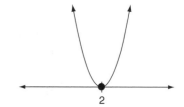

The curve is touching or below the x-axis ($y \leq 0$) for $x = 2$.

8. $4 + 2x^2 \leq 0$

Since $a = 2$, the parabola opens up.

$$4 + 2x^2 = 0$$
$$2x^2 = -4$$
$$x^2 = -2$$

There is no x-intercept.

The curve is never touching or below the x-axis ($y \leq 0$), so there is no solution.

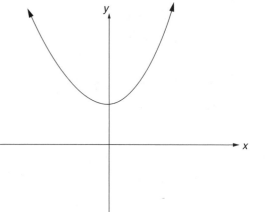

9. $9x^2 - 9x \leq 0$

Let $y = 9x^2 - 9x$ and graph. Since $a = 9$, the parabola opens up.

$$9x^2 - 9x = 0$$
$$9x(x - 1) = 0$$
$$x = 0 \quad \text{or} \quad x = 1$$

The x-intercepts are 0 and 1.

Make a rough sketch.

The curve is touching or below the x-axis ($y \leq 0$) for $0 \leq x \leq 1$.

10. $25 - x^2 \geq 0$

Since $a = -1$, the parabola opens down.

$$25 - x^2 = 0$$
$$(5 - x)(5 + x) = 0$$

$x = 5$ or $x = -5$

The x-intercepts are -5 and 5.

The curve is touching or above the x-axis
($y \geq 0$) for $-5 \leq x \leq 5$.

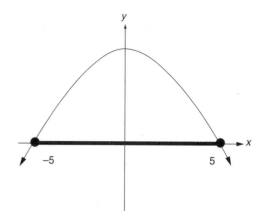

11. $x^2 + 1 \geq 0$

Let $y = x^2 + 1$ and graph. Since $a = 1$, the parabola opens up.

$$x^2 + 1 = 0$$
$$x^2 = -1$$

There is no x-intercept.

Make a rough sketch.

The curve is touching or above the x-axis ($y \leq 0$) for all values of x.

The solution is the set of all real numbers.

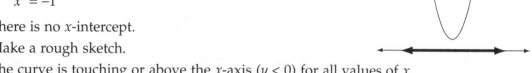

12. $x^2 + 6x + 9 > 0$

Since $a = 1$, the parabola opens
up.

$$x^2 + 6x + 9 = 0$$
$$(x + 3)(x + 3) = 0$$

$x = -3$ or $x = -3$

The x-intercept is -3.
$x < -3$ or $x > -3$, in other
words, all reals except -3.

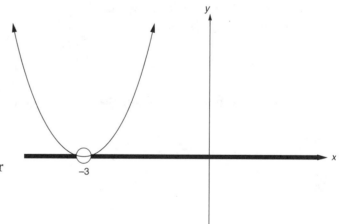

13. $-x^2 + 2x + 15 \leq 0$

Let $y = -x^2 + 2x + 15$ and graph. Since $a = -1$, the parabola opens down.

$$-x^2 + 2x + 15 = 0$$
$$x^2 - 2x - 15 = 0$$
$$(x - 5)(x + 3) = 0$$

$x = 5$ or $x = -3$

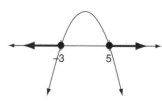

The x-intercepts are -3 and 5.

Make a rough sketch.

The curve is touching or below the x-axis ($y \leq 0$) for $x \leq -3$ or $x \geq 5$.

14. $2x^2 + 1 \leq 0$

Let $y = 2x^2 + 1$ and graph. Since $a = 2$, the parabola opens up.

$2x^2 + 1 = 0$

$\quad 2x^2 = -1$

$\quad x^2 = \dfrac{-1}{2}$

There is no x-intercept.

Make a rough sketch.

The curve is never touching or below the x-axis ($y \leq 0$).

There is no solution.

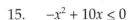

15. $-x^2 + 10x \leq 0$

Since $a = -1$, the parabola opens down.

$-x^2 + 10x = 0$

$\quad x^2 - 10x = 0$

$\quad x(x - 10) = 0$

$\quad x = 0 \quad$ or $\quad x = 10$

The x-intercepts are 0 and 10.

The curve is touching or below the x-axis ($y \leq 0$) for $x \leq 0$ or $x \geq 10$.

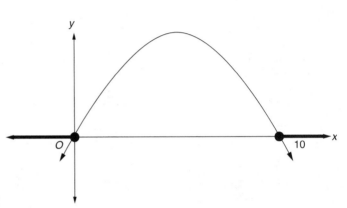

16. $2x^2 + 3x - 5 \leq 0$

Let $y = 2x^2 + 3x - 5$ and graph. Since $a = 2$, the parabola opens up.

$\qquad 2x^2 + 3x - 5 = 0$

$\qquad 2x^2 + 5x - 2x - 5 = 0$

$\quad x(2x + 5) - 1(2x + 5) = 0$

$\qquad (2x + 5)(x - 1) = 0$

$x = -\dfrac{5}{2} \quad$ or $\quad x = 1$

The x-intercepts are $-\dfrac{5}{2}$ and 1.

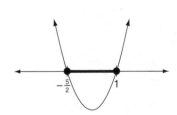

Make a rough sketch.

The curve is touching or below the x-axis ($y \leq 0$) for $\dfrac{-5}{2} \leq x \leq 1$.

17. $-6x^2 - x + 2 \geq 0$

Let $y = -6x^2 - x + 2$ and graph. Since $a = -6$, the parabola opens down.

$\qquad -6x^2 - x + 2 = 0$

$\qquad 6x^2 + x - 2 = 0$

$\qquad 6x^2 + 4x - 3x - 2 = 0$

$$2x(3x + 2) - 1(3x + 2) = 0$$
$$(3x + 2)(2x - 1) = 0$$
$$x = -\frac{2}{3} \quad \text{or} \quad x = \frac{1}{2}$$

The x-intercepts are $-\frac{2}{3}$ and $\frac{1}{2}$.

Make a rough sketch.

The curve is touching or above the x-axis $(y \geq 0)$ for $\frac{-2}{3} \leq x \leq \frac{1}{2}$.

UNIT 30

1. $|2x - 5| = 21$

    $$2x - 5 = 21 \quad \text{or} \quad -(2x - 5) = 21$$
    $$2x = 26 \qquad\qquad -2x + 5 = 21$$
    $$x = 13 \qquad\qquad\qquad -2x = 16$$
    $$\qquad\qquad\qquad\qquad x = -8$$

    Both check. The solutions are $x = -8$ or $x = 13$.

2. $|x - 7| + 2x = 5$

    $$x - 7 + 2x = 5 \quad \text{or} \quad -(x - 7) + 2x = 5$$
    $$3x = 12 \qquad\qquad -x + 7 + 2x = 5$$
    $$x = 4 \qquad\qquad\qquad\qquad x = -2$$

    Only $x = -2$ checks. The solution is $x = -2$.

3. $|5 - 2x| = -3$

    $$5 - 2x = -3 \quad \text{or} \quad -(5 - 2x) = -3$$
    $$-2x = -8 \qquad\qquad -5 + 2x = -3$$
    $$x = 4 \qquad\qquad\qquad 2x = 2$$
    $$\qquad\qquad\qquad\qquad x = 1$$

    Neither solution checks. There is no solution.

4. $3x - |x + 5| = 9$

    $$3x - (x + 5) = 9 \quad \text{or} \quad 3x - -(x + 5) = 9$$
    $$3x - x - 5 = 9 \qquad\qquad 3x + x + 5 = 9$$
    $$2x = 14 \qquad\qquad\qquad 4x = 4$$
    $$x = 7 \qquad\qquad\qquad\qquad x = 1$$

    Only $x = 7$ checks. The solution is $x = 7$.

5. $y = |x - 4| - 2$

    The line of symmetry is $x = 4$.

    Table of values

x	y
2	0
3	-1
4	-2
5	-1
6	0

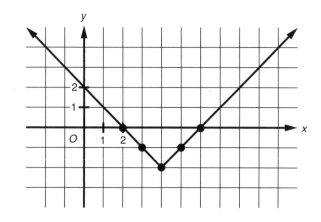

6. $y = -|x + 3| + 5$

    The line of symmetry is $x = -3$.

    Table of values

x	y
-5	3
-4	4
-3	5
-2	4
-1	3

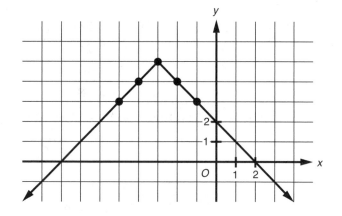

7. $y = \dfrac{1}{2}|x - 2|$

    The line of symmetry is $x = 2$.

    Table of values

x	y
-2	2
0	1
2	0
4	1
6	2

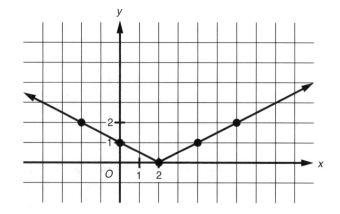

8. $y = -|2x + 1| + 7$

    The line of symmetry is $x = \dfrac{-1}{2}$.

    Table of values

x	y
-3	2
-2	4
-1	6
$-1/2$	7
0	6
1	4
2	2

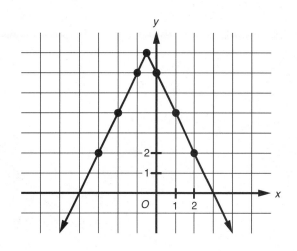

9. $|3x - 7| \le 8$

    $$3x - 7 = 8 \qquad \text{or} \qquad -(3x - 7) = 8$$
    $$3x = 15 \qquad\qquad\qquad -3x + 7 = 8$$
    $$x = 5 \qquad\qquad\qquad\qquad -3x = 1$$
    $$x = \dfrac{-1}{3}$$

    The solutions are $\dfrac{-1}{3} \le x \le 5$.

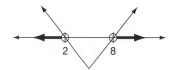

10. $|5 - x| > 3$

    $$5 - x = 3 \qquad \text{or} \qquad -(5 - x) = 3$$
    $$-x = -2 \qquad\qquad\qquad -5 + x = 3$$
    $$x = 2 \qquad\qquad\qquad\qquad x = 8$$

    The solutions are $x < 2$ or $x > 8$.

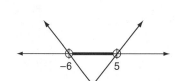

11. $|2x + 1| < 11$

    $$2x + 1 = 11 \qquad \text{or} \qquad -(2x + 1) = 11$$
    $$2x = 10 \qquad\qquad\qquad -2x - 1 = 11$$
    $$x = 5 \qquad\qquad\qquad\qquad -2x = 12$$
    $$x = -6$$

    The solutions are $-6 < x < 5$.

12. $-2|x + 3| + 4 \ge 0$

    $$-2(x + 3) + 4 = 0 \qquad \text{or} \qquad 2(x + 3) + 4 = 0$$
    $$-2x - 6 + 4 = 0 \qquad\qquad\qquad 2x + 6 + 4 = 0$$
    $$-2x - 2 = 0 \qquad\qquad\qquad\qquad 2x + 10 = 0$$
    $$-2x = 2 \qquad\qquad\qquad\qquad\qquad 2x = -10$$
    $$x = -1 \qquad\qquad\qquad\qquad\qquad x = -5$$

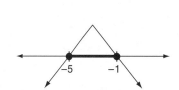

    The solutions are $-5 \le x \le -1$.

UNIT 31

1. $\log_3 81 = x$
   $3^x = 81$
   $= 3^4$
   $x = 4$

2. $\log_5 125 = x$
   $5^x = 125$
   $= 5^3$
   $x = 3$

3. $\log_7\left(\dfrac{1}{7}\right) = x$
   $7^x = \dfrac{1}{7}$
   $= 7^{-1}$
   $x = -1$

4. $\log 1 = x$
   $10^x = 1$
   $x = 0$

5. $\log_3 x = 2$
   $3^2 = x$
   $9 = x$

6. $\log_7 x = 0$
   $7^0 = x$
   $1 = x$

7. $\log_9 x = \dfrac{1}{2}$
   $9^{1/2} = x$
   $\sqrt{9} = x$
   $3 = x$

8. $\log_x 27 = 3$
   $x^3 = 27$
   $= 3^3$
   $x = 3$

9. $\log_x 49 = 2$
   $x^2 = 49$
   $= 7^2$
   $x = 7$

10. $\log_x 121 = 2$
    $x^2 = 121$
    $x = 11$

11. $\log x^5 = 5 \log x$

12. $\log 2xy^3$
    $\log 2 + \log x + \log y^3$
    $\log 2 + \log x + 3 \log y$

13. $\log \dfrac{x^2}{y}$
    $\log x^2 - \log y$
    $2 \log x - \log y$

14. $\log \dfrac{x}{yz}$
    $\log x - \log yz$
    $\log x - (\log y + \log z)$
    $\log x - \log y - \log z$

15. $\log \sqrt{x^3 y}$
    $\log(x^3 y)^{1/2}$
    $\dfrac{1}{2} \log x^3 y$
    $\dfrac{1}{2}(\log x^3 + \log y)$
    $\dfrac{1}{2}(3 \log x + \log y)$
    $\dfrac{3}{2} \log x + \dfrac{1}{2} \log y$

UNIT 32

1.

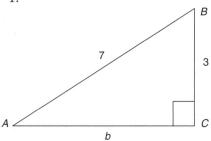

$$c^2 = a^2 + b^2$$
$$7^2 = 3^2 + b^2$$
$$49 = 9 + b^2$$
$$40 = b^2$$
$$b^2 = 40$$
$$b = \sqrt{40}$$
$$b = 2\sqrt{10}$$

2.

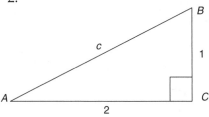

$$c^2 = a^2 + b^2$$
$$c^2 = 1^2 + 2^2$$
$$c^2 = 1 + 4$$
$$c^2 = 5$$
$$c = \sqrt{5}$$

3.

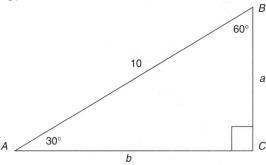

$$a = \frac{c}{2}$$
$$a = \frac{10}{2}$$
$$a = 5$$
$$b = \frac{c}{2}\sqrt{3}$$
$$b = \frac{10}{2}\sqrt{3}$$
$$b = 5\sqrt{3}$$

4.

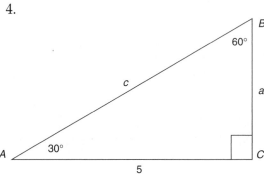

$$b = \frac{c}{2}\sqrt{3}$$
$$5 = \frac{c}{2}\sqrt{3}$$
$$10 = c\sqrt{3}$$
$$\frac{10}{\sqrt{3}} = c \quad \text{or} \quad c = \frac{10\sqrt{3}}{3}$$
$$a = \frac{c}{2}$$
$$a = \frac{10}{\sqrt{3}} \cdot \frac{1}{2}$$
$$a = \frac{5}{\sqrt{3}} \quad \text{or} \quad a = \frac{5\sqrt{3}}{3}$$

5.

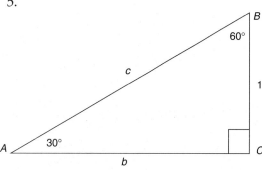

$$a = \frac{c}{2}$$

$$1 = \frac{c}{2}$$

$$2 = c$$

$$b = \frac{c}{2}\sqrt{3}$$

$$b = \frac{2}{2}\sqrt{3}$$

$$b = \sqrt{3}$$

6.

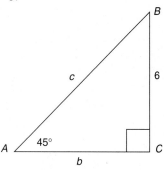

$$a = b$$

$$6 = b$$

$$a = \frac{c}{2}\sqrt{2}$$

$$6 = \frac{c}{2}\sqrt{2}$$

$$12 = c\sqrt{2}$$

$$\frac{12}{\sqrt{2}} = c \quad \text{or} \quad c = 6\sqrt{2}$$

7.

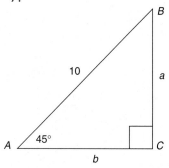

$$a = \frac{c}{2}\sqrt{2}$$

$$a = \frac{10}{2}\sqrt{2}$$

$$a = 5\sqrt{2}$$

$$b = 5\sqrt{2}$$

8.

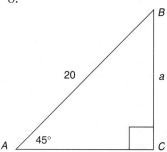

$$a = \frac{c}{2}\sqrt{2}$$

$$a = \frac{20}{2}\sqrt{2}$$

$$a = 10\sqrt{2}$$

The ladder reaches up $10\sqrt{2}$ ft.

9.

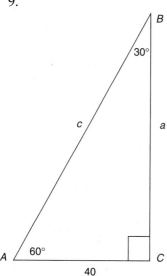

$$b = \frac{c}{2}$$

$$40 = \frac{c}{2}$$

$$80 = c$$

$$a = \frac{c}{2}\sqrt{3}$$

$$a = \frac{80}{2}\sqrt{3}$$

$$a = 40\sqrt{3}$$

The building is $40\sqrt{3}$ ft high.

10.

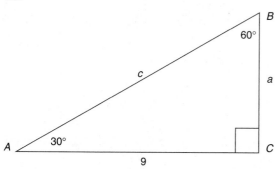

$$b = \frac{c}{2}\sqrt{3}$$

$$9 = \frac{c}{2}\sqrt{3}$$

$$18 = c\sqrt{3}$$

$$\frac{18}{\sqrt{3}} = c \quad \text{or} \quad c = 6\sqrt{3}$$

$$a = \frac{c}{2}$$

$$a = \frac{18}{\sqrt{3}} \cdot \frac{1}{2}$$

$$a = \frac{9}{\sqrt{3}} \quad \text{or} \quad a = 3\sqrt{3}$$

The observer is $\frac{9}{\sqrt{3}}$ or $3\sqrt{3}$ ft tall.

Index

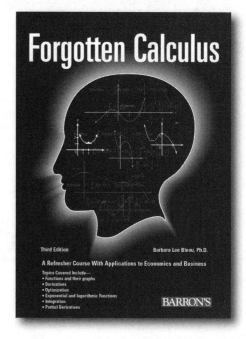